Maxwell's Equations and their applications

Student Monographs in Physics

Series Editor: Professor Douglas F Brewer
Professor of Experimental Physics, University of Sussex

Other books in the series:

Microcomputers
D G C Jones

Oscillations and Waves
R Buckley

Fourier Transforms in Physics
D C Champeney

Maxwell's Equations and their Applications

E G Thomas

Department of Physics, University of Leicester

A J Meadows

Department of Astronomy, University of Leicester

Adam Hilger Ltd, Bristol and Boston

British Library Cataloguing in Publication Data

Thomas, E. G.

Maxwell's equations and their applications.—(Student monographs in physics series)

1. Electromagnetic theory

I. Title II. Meadows, A. J. III. Series

530.1'41 QC670

ISBN 0-85274-778-0

Published by Adam Hilger Ltd
Techno House, Redcliffe Way, Bristol BS1 6NX, England
PO Box 230, Accord, MA 02018, USA

Printed in Great Britain by Page Bros (Norwich) Ltd

Contents

Preface

This book aims to give a brief, but self-contained, introduction to the derivation and use of Maxwell's equations. Much of electromagnetism—one of the major branches of physics—is based on these equations. They therefore form an essential part of all physics courses. At the same time, many undergraduate students seem to find their application difficult. For this reason, although all the numerous texts on electricity and magnetism discuss Maxwell's equations, we believe that they deserve a separate, close examination on their own. Our aim in this text is to explore the physical meaning of the equations and to bring out the relationship between this physical content and its mathematical representation. We have found that undergraduates experience particular difficulty in making this connection. The text introduces the standard vector operator notation, as this is essential to the subject, but otherwise the mathematical requirements have been kept to a minimum.

We start in the first chapter by defining electric and magnetic fields. The idea of 'lines of force' is introduced from the beginning as an aid to visualising the fields. Maxwell's equations are derived in considerable detail, along with two supplementary equations. The chapter ends with a short discussion of electromagnetic units.

The second chapter looks at applications of Maxwell's equations in the simplest case (in a vacuum). We start with problems where the conditions are steady state, for which only two of Maxwell's equations are required. The full set of equations is then employed to demonstrate the existence of electromagnetic waves in free space. After an examination of the propagation of plane electromagnetic waves, the chapter ends with a discussion of energy and momentum conservation in electromagnetism.

The final chapter considers how Maxwell's equations can be applied to situations where bulk matter is present, rather than a vacuum. We start by looking at the implications for the propagation of electromagnetic waves and end with a problem which requires the full array of Maxwell's equations.

We have assumed throughout this book that the reader already has some knowledge of elementary vector analysis. In case of doubt, there are a large number of introductory texts that can be consulted. Noteworthy amongst these is the book *Vector Analysis* by M R Spiegel (1977, McGraw-Hill). We expect the reader to have some acquaintance with vector addition and with scalar and vector products. For example, we use the identity $\boldsymbol{A} \wedge (\boldsymbol{B} \wedge \boldsymbol{C}) = \boldsymbol{B}(\boldsymbol{A} \cdot \boldsymbol{C}) - \boldsymbol{C}(\boldsymbol{A} \cdot \boldsymbol{B})$. (For a proof of this see, e.g. Spiegel p 28.) The reader will also need to be familiar with the definitions and meanings of the 'grad', 'div' and 'curl' operators and their representations in cartesian coordinates. (A treatment of these topics can be found in Spiegel, Chapter 4.) Finally, we use in the text the integral vector relationships known as 'Gauss' divergence theorem' and 'Stokes' theorem'. (These are dealt with in Spiegel, Chapter 6.)

We would like to thank Professor D F Brewer, Professor E A Davis and Mr N Hankins for their comments.

Maxwell's Equations—the Simplest Form

1

1.1 Introduction

A century ago, the Scottish scientist James Clerk Maxwell brought together and extended four basic laws dealing with electromagnetism. These laws govern the interaction of bodies which are magnetic or electrically charged or both. The bodies and their charges may either be stationary or in motion. As we shall see, the number four is significant in that it reflects a degree of symmetry in the nature of electrical and magnetic forces; although the laws are based on various kinds of observational data, they are interrelated. The mathematical equations which describe these basic laws take their simplest form when the interacting bodies are situated in a vacuum. We shall assume that this is the case in deriving Maxwell's four equations in this chapter.

There are two ways of thinking about the interaction of two objects separated in space from each other. In one, attention is concentrated only on the two objects, and an appropriate force equation (depending on their distance apart etc) is used to determine the effect of one body on the other. At first sight, this seems the obvious way of doing things, but in electromagnetism, it often provides less insight into what is happening than an alternative approach, in which more attention is paid to what is happening in the space between the two bodies. We can suppose that this space is filled with a 'field' whose properties determine how the bodies interact.

The difference in the two approaches can be seen if we consider, for example, the interaction between two stationary electrons. We start with the experimental result which is usually called Coulomb's inverse-square law (after the French scientist who established it a couple of centuries ago). This states that an electrical charge q_1 exerts a force F on another electrical charge q_2 separated from it by a distance r, where

$$F = \frac{1}{4\pi\varepsilon_0} \frac{q_1 q_2}{r^2}.$$

(The factor $(1/4\pi\varepsilon_0)$ is a constant introduced to get the units right; we will discuss this further at the end of the chapter.)

Since force is a vector quantity, Coulomb's law is often rewritten in the form

$$\boldsymbol{F}=\frac{1}{4\pi\varepsilon_0}\frac{q_1q_2}{r^3}\,\boldsymbol{r}. \tag{1.1}$$

This equation indicates that the force acts along the line joining the two bodies. For two electrons, the charges are equal in magnitude and both are negative. Instead of q_1q_2, we can therefore write $(-q)(-q)=q^2$. This means that the force is positive and acts in the direction of increasing r, measured from either of the point charges. In other words, we are dealing with a repulsive force.

If we add more electrons to our original two, the picture obviously becomes more complicated. However, we find experimentally that this does not affect the interaction between the original two electrons, which can still be described in terms of the same law. What this means is that the interactions of one charged body with any number of others can be added together. Since we are dealing with a vector equation, the addition has to be done vectorially. Figure 1.1 shows an example of this for three stationary electrons.

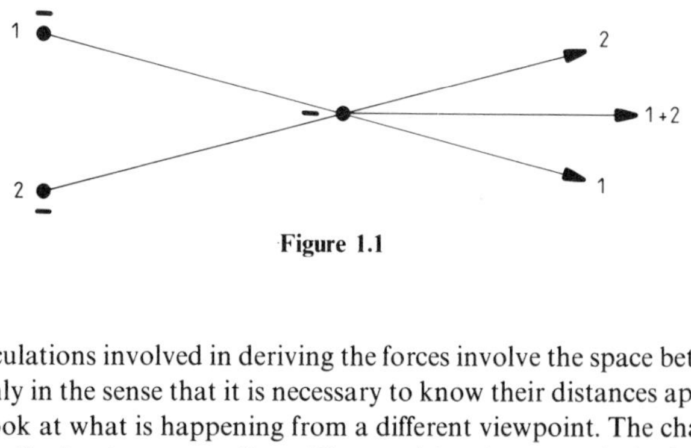

Figure 1.1

The calculations involved in deriving the forces involve the space between the charges only in the sense that it is necessary to know their distances apart. Now we shall look at what is happening from a different viewpoint. The charge q_1 is obviously affecting its surroundings. The repulsion (or attraction) of q_2 reveals this influence, but it would presumably still be there even if q_2 were taken away. We can say that q_1 is producing an electric field $\boldsymbol{E}$ in the space around it; when q_2 is introduced, the field in its vicinity acts on it so that it experiences a force $\boldsymbol{F}=q_2\boldsymbol{E}$. (Note that, since force is a vector, and electrical charge is a scalar, the field must be a vector acting in the same direction as the force if q_2 is positive, and in the opposite direction if q_2 is negative.) This can be reconciled with Coulomb's law, if

$$\boldsymbol{E}=\frac{1}{4\pi\varepsilon_0}\frac{q_1}{r^3}\,\boldsymbol{r}. \tag{1.2}$$

We can formally define a 'field' as a representation of the way in which some distributed quantity (in this case, $\boldsymbol{E}$) varies with position. But the significant point is that the word 'field' transfers our attention from the electrical charges to the space around them. (In the same way, talking of a football field concentrates our attention on the place where football is played, rather than on the players.) The next step is to try to construct a picture of what a field is like. Consider a bar magnet placed on a sheet of cardboard on which iron filings have been scattered. If the cardboard is tapped gently, the filings tend to line themselves up in loops which join the north and south poles of the magnet as shown in figure 1.2. This experiment suggests that one possible representation of a field is by a series of imaginary lines extending through the space around magnets or electrical charges. Since a field exerts a force, we usually call these 'lines of force'. Correspondingly, a magnetic pole can be defined as the point from which lines of force diverge (for a north pole) or converge (for a south pole).

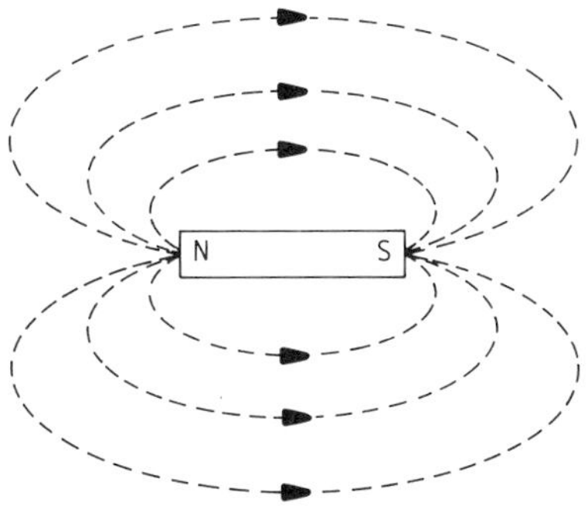

Figure 1.2

A major problem in thinking about electromagnetism as a unified whole is the fundamental difference that exists between electricity and magnetism. Whereas isolated electrical charges can be easily produced in the laboratory and experimented on there, isolated magnetic poles have never so far been identified. Consequently magnetic 'poles' occur only in theory: they are not something that can be experimented on directly in the laboratory. However, it is useful in developing Maxwell's equations to suppose there is a symmetry between electricity and magnetism such that both can be represented by poles. We shall do so here.

For our picture to be put to physical use, it must be made quantitative. The electric field has two properties which must be represented—direction and magnitude (usually called 'intensity'). We can satisfy the former by saying that the direction of $\boldsymbol{E}$ at any point corresponds to the direction of the line of force passing through that point. (This definition means that two lines of force cannot intersect: otherwise the field would have two different directions simultaneously.) To avoid ambiguity we define the direction of lines of force as being *away* from north magnetic poles or positive electrical charges and *towards* south magnetic poles or negative electrical charges.

The intensity of the field is defined as the number of lines of force passing through a unit area at right angles to the direction of the lines; the more intense the field, the higher the concentration of lines of force used to represent it. Since lines of force only originate and end on magnetic poles or electrical charges, the number present for a given magnet or charged body must be constant. This does not mean, however, that the intensity cannot vary with position. For example, consider an electrical charge q. According to our definition, the number of lines of force it produces per unit area is given by $E=(1/4\pi\varepsilon_0)q/r^2$, so the intensity varies inversely as the square of the distance. In figure 1.3, suppose we draw an imaginary sphere 1 round this charge at a distance r and ask how many lines penetrate it. The answer is obviously $(1/4\pi\varepsilon_0)(q/r^2)\times 4\pi r^2=q/\varepsilon_0$. This is independent of distance, so the total number of lines of force must be constant all the way from the charge to infinity. (The same result can be obtained immediately by drawing another sphere 2 around the charge, and comparing with sphere 1.) Note that this argument only works if the force falls off as the inverse square of the distance, which is true for the point charges we have been considering.

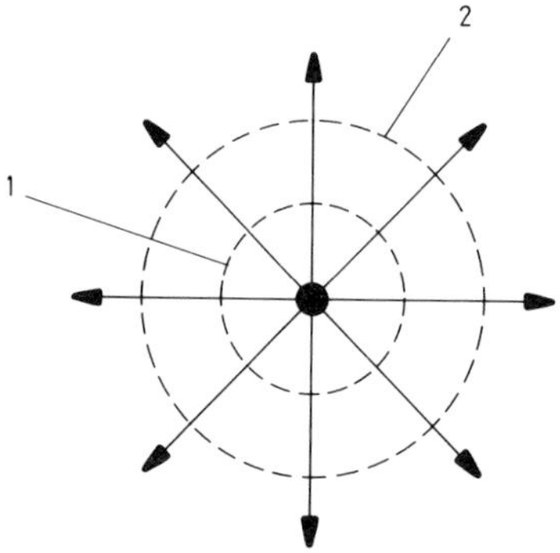

Figure 1.3

The intensity of a field may be very small, much less than unity. This raises the question: how can you have a fraction of a line of force? The answer is that fractions only arise because of the units we use. If we linked one line of force to each elementary unit of charge (such as an electron or magnetic pole)—as we could do—the problem would not arise. In principle, we could replace each line of force by a 'tube' composed of a large bundle of individual lines. A fractional value for the intensity of the field would then simply indicate the proportion of lines of force in the tube as compared with some arbitrary standard number.

If we work out where the lines of force go for a large number of electrical and magnetic configurations, we find that their layout can always be explained by attributing two properties to lines of force in addition to those we have mentioned. The first is that they are in tension along their lengths; the second is that they repel each other. For example, suppose we plot—perhaps using iron filings—the directions of lines of force between the north pole of one magnet and the south pole of another. The result is shown in figure 1.4. The tension along the lines is evident from the tendency of the two poles to pull themselves together. However, the lines do not go straight across from one pole to the other, as would be expected if they were simply under tension. Instead, they bulge out sideways, reflecting their mutual repulsion at right angles to each other. Since the number of lines per unit area measures intensity, this diagram helps us to understand why there is a smaller, but significant, magnetic field outside the immediate region of the poles. *Lines of force cannot be used to interpret everything that happens in electromagnetism*—for this we must turn to the basic equations—but they can often provide insight into what is happening. We will therefore keep them in mind while we derive Maxwell's equations.

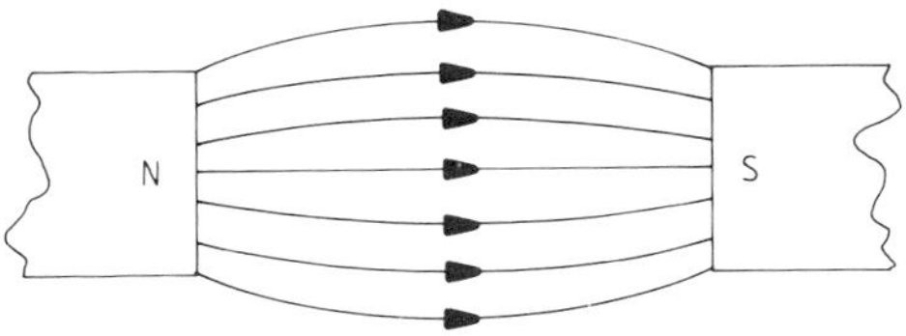

Figure 1.4

The four equations can be divided up in the following way.

1 Electricity Static	2 Magnetism Static
3 Electricity Varying	4 Magnetism Varying

We will consider them in the order in which they are numbered in this diagram.

1.2 Electrostatics

We have seen that the number of lines of force N leaving an electrical charge is given by $N = q/\varepsilon_0$. Suppose that, instead of a single charge, we have an array of charges $q_1, \ldots, q_n$. We draw an imaginary closed surface round the volume of space they occupy, and remember that their effects can be added together. This means that the number of lines of force penetrating the surface is $N = (1/\varepsilon_0) \sum_{i=1}^{n} q_i$. In doing the addition, lines going outward through the surface are counted as positive and lines going inwards as negative. So, if there are equal numbers of positive and negative charges within the surface, the net number of lines of force through the surface will be zero.

We have defined the number of lines of force per unit area (normal to the lines) as $\boldsymbol{E}$, the intensity of the field. We can therefore write $N=\int_A \boldsymbol{E}\cdot \mathrm{d}\boldsymbol{A}$. Here $\mathrm{d}\boldsymbol{A}$ is an element of area on our imaginary surface. It is written as a vector because each small area can be represented by an arrow of length proportional to the area and direction perpendicular to its surface (see figure 1.5(*a*)). We have taken the scalar product of $\boldsymbol{E}$ and $\mathrm{d}\boldsymbol{A}$ to ensure that we are measuring the number of lines of force through an area at right angles to the field ($=E\cos\theta\, \mathrm{d}A$; see figure 1.5(*b*)). The symbol $\int_A$ means that the integration is carried out over the entire surface enclosing the charges. $\int_A \boldsymbol{E}\cdot \mathrm{d}A$ is called the flux of $\boldsymbol{E}$ across the surface A.

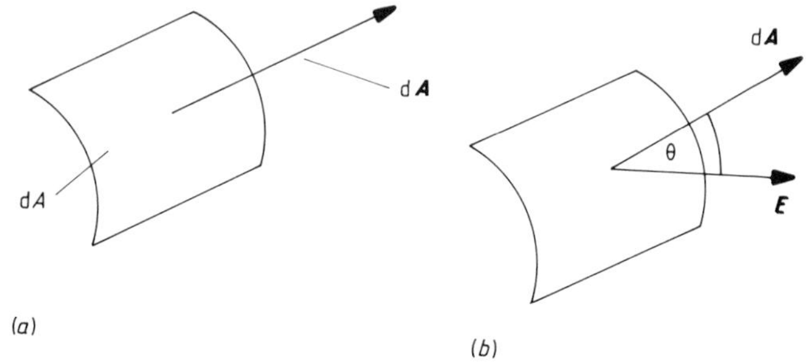

Figure 1.5

We can now write $N=\int_A \boldsymbol{E}\cdot \mathrm{d}\boldsymbol{A}=(1/\varepsilon_0)\sum_{i=1}^{n} q_i$. This result is usually referred to as Gauss's law, after the German mathematician of the early nineteenth century. More generally, the electrical charge need not be present simply as a set of separate sources. Instead, it may be distributed throughout the volume contained by the surface, perhaps varying in amount from place to place. We can describe such a distribution in terms of an electrical charge density per unit volume, ρ. In this case, we must replace $\sum_{i=1}^{n} q_i$ by an integral $\int_V \rho\, \mathrm{d}V$. Here, $\mathrm{d}V$ is an element of volume and the V beside the integral sign indicates that the integration is taken over the whole volume. Gauss's law therefore becomes

$$\int_A \boldsymbol{E}\cdot \mathrm{d}\boldsymbol{A}=\frac{1}{\varepsilon_0}\int_V \rho\, \mathrm{d}V. \tag{1.3}$$

The left-hand side of this equation can also be rewritten. The divergence of a vector can be defined at any point in space as

$$\operatorname{div} \boldsymbol{E}=\lim_{V\to 0}\left(\int_A \boldsymbol{E}\cdot \mathrm{d}\boldsymbol{A}\right)\frac{1}{V}.$$

If we take the specified limit, this is equivalent to saying that

$$\int_V \operatorname{div} \boldsymbol{E}\, \mathrm{d}V=\int_A \boldsymbol{E}\cdot \mathrm{d}A. \tag{1.4}$$

This mathematical result is sometimes known as Gauss's *theorem* (not law).

The physical significance of the divergence is that it measures the net amount of the quantity under consideration which enters or leaves each small volume of space. For an electrical field, the quantity concerned is the number of lines of force. (The definition of divergence shows that it is a scalar quantity, so that it is appropriate to represent it by a number.) Consider again an imaginary surface in a region of space containing electrical fields. If the surface does not enclose any charge then the number of lines entering the volume is equal to the number of lines leaving the volume, this is because lines of force only start and end on charges (see figure 1.6(*a*)). Thus $\int_A \boldsymbol{E}\cdot \mathrm{d}\boldsymbol{A}$ is zero, and from equation (1.4) $\int_V \mathrm{div}\, \boldsymbol{E}\,\mathrm{d}V$ is therefore also equal to zero. On the other hand, a surface which does enclose a net charge will have a net flux of lines of force across its surface as can be seen from figure 1.6(*b*). Clearly $\int_V \mathrm{div}\, \boldsymbol{E}\,\mathrm{d}V$ is non-zero in this case.

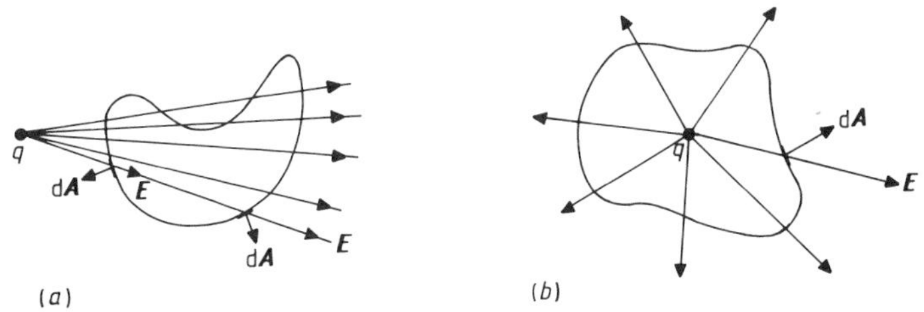

Figure 1.6

If we combine the two equations for $\int_A \boldsymbol{E}\cdot \mathrm{d}\boldsymbol{A}$, we have $\int_V \mathrm{div}\, \boldsymbol{E}\,\mathrm{d}V = (1/\varepsilon_0)\int_V \rho\,\mathrm{d}V$. Differentiating both sides leads immediately to the equation $\mathrm{div}\, \boldsymbol{E} = \rho/\varepsilon_0$. (What this means physically is that, since $\mathrm{d}V$ appears on both sides of the equation, the relationship between div $\boldsymbol{E}$ and ρ must hold for any arbitrary point in space.) This is actually a statement of the first of Maxwell's equations, but we will express it in a rather different form.

We can write the divergence of $\boldsymbol{E}$ in a number of equivalent ways. In cartesian coordinates, we have $\mathrm{div}\, \boldsymbol{E} = \partial E_x/\partial x + \partial E_y/\partial y + \partial E_z/\partial z$. Notice that we are using partial differentiation allows us to concentrate on, say the gradient of E in the x direction, and ignore changes in the y and z directions. The form we will partial differentiation allows us to concentrate on, say the gradient of E in the x direction, and ignore changes in the y and z directions. The form we will preferentially use for the divergence is $\mathrm{div}\, \boldsymbol{E} = \nabla \cdot \boldsymbol{E}$. The vector operator ∇ (sometimes called 'nabla', and sometimes 'del'—short for 'delta') is defined as $\boldsymbol{i}\partial/\partial x + \boldsymbol{j}\partial/\partial y + \boldsymbol{k}\partial/\partial z$ in cartesian coordinates (where $\boldsymbol{i}, \boldsymbol{j}$ and $\boldsymbol{k}$ are the unit vectors along the x, y and z axes, as shown in figure 1.7(*b*)). We can therefore write

$$\nabla \cdot \boldsymbol{E} = \left(\boldsymbol{i}\frac{\partial}{\partial x} + \boldsymbol{j}\frac{\partial}{\partial y} + \boldsymbol{k}\frac{\partial}{\partial z}\right)\cdot(\boldsymbol{i}E_x + \boldsymbol{j}E_y + \boldsymbol{k}E_z) = \frac{\partial E_x}{\partial x} + \frac{\partial E_y}{\partial y} + \frac{\partial E_z}{\partial z}$$

corresponding, as it should, to the representation of div $\boldsymbol{E}$ in cartesian coordinates.

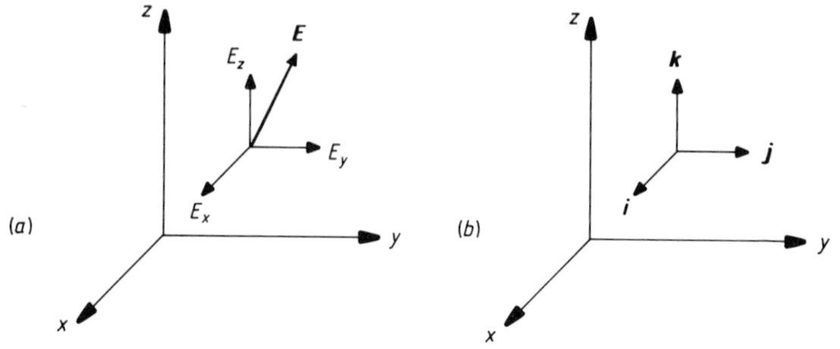

Figure 1.7

We finally write the first Maxwell equation as

$$\nabla \cdot \boldsymbol{E} = \rho / \varepsilon_0. \tag{1.5}$$

1.3 Magnetostatics

The second of Maxwell's equations, dealing with static magnetic fields, now follows very quickly. We define a magnetic 'intensity' $\boldsymbol{B}$ corresponding to the electric intensity $\boldsymbol{E}$. (We have written 'intensity', even though $\boldsymbol{B}$ is usually called the magnetic 'induction'.) Experiments indicate that the properties attributed to electrical charges and, in particular, the representation of the field by lines of force, can be applied equally to magnetic poles. This means that we can write for the number of lines of force N:

$$N = \int_A \boldsymbol{B} \cdot \mathrm{d}\boldsymbol{A} = \int_V \nabla \cdot \boldsymbol{B} \, \mathrm{d}V. \tag{1.6}$$

Suppose we now draw our imaginary surface round any distribution of magnets in space. Each magnet consists of a north and a south pole, so the number of lines of force going outwards through the surface from the north poles must be exactly balanced by the number coming inwards through the surface to the south poles. It must therefore always be true that the net number of lines is zero, and

$$N = \int_V \nabla \cdot \boldsymbol{B} \, \mathrm{d}V = 0.$$

Differentiating this equation, we reach the second of Maxwell's equations

$$\nabla \cdot \boldsymbol{B} = 0. \qquad (1.7)$$

A comparison with the corresponding equation (1.5) for $\boldsymbol{E}$ underlines the difference between electricity and magnetism. Isolated charges can exist, but magnetic monopoles (as they are called) cannot; at least, no search for isolated magnetic poles has yet been successful.

We will now look at what happens when a constant electrical current flows through a conductor. In the early nineteenth century, two French scientists found a relationship between the current flowing along a wire and the magnetic field it produces. This field due to a current i flowing along an element of the wire $\mathrm{d}\boldsymbol{s}$ is given by

$$\mathrm{d}\boldsymbol{B} = \frac{\mu_0 i}{4\pi r^3}(\mathrm{d}\boldsymbol{s} \wedge \boldsymbol{r}). \qquad (1.8)$$

(This equation is known as the Biot–Savart law after its discoverers.) The relationship tells us two important things about the magnetic field. First, since we have $\boldsymbol{r}$ in the numerator and r^3 in the denominator, we are dealing once again with an inverse-square law. Second, the direction of the vector $\mathrm{d}\boldsymbol{B}$ is given by the vector product of $\mathrm{d}\boldsymbol{s}$ and $\boldsymbol{r}$. This means that the magnetic field is at right angles to both $\mathrm{d}\boldsymbol{s}$ and $\boldsymbol{r}$. In addition, we have introduced a constant μ_0 which depends on the units we are using. It parallels, for the magnetic field, the constant ε_0 which we have used for electric fields.

We will now apply this law to a very long, straight wire carrying a current i in order to determine the magnitude of $\mathrm{d}\boldsymbol{B}$. In this case (see figure 1.8) $\mathrm{d}\boldsymbol{s} \wedge \boldsymbol{r} = r \sin\theta \, \mathrm{d}s$. So $\mathrm{d}B = \mu_0 i \sin\theta \, \mathrm{d}s / 4\pi r^2$. We also have $p/r = \sin\theta$ (where p is the perpendicular distance from the point where we are measuring the field to the wire). In addition, $r\,\mathrm{d}\theta/\mathrm{d}s = \sin\theta$. Eliminating r between the last two equations gives $\mathrm{d}s = p\,\mathrm{d}\theta/\sin^2\theta$. Putting this back into our equation for $\mathrm{d}B$ and substituting again for r, we finally have

$$\mathrm{d}B = \frac{\mu_0 i \sin\theta \sin^2\theta \, p \, \mathrm{d}\theta}{4\pi p^2 \sin^2\theta} = \frac{\mu_0 i \sin\theta \, \mathrm{d}\theta}{4\pi p}.$$

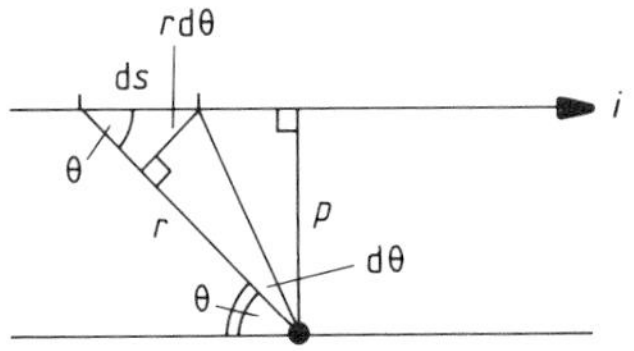

Figure 1.8

For a very long wire we can assume that the value of θ corresponding to one end of it will be zero, while the other end will be at approximately 180° ($=\pi$ radians). The total magnetic field at our point of measurement will therefore be

$$B=\frac{\mu_0 i}{4\pi p}\int_0^{\pi}\sin\theta\,\mathrm{d}\theta=\frac{\mu_0 i}{2\pi p}. \tag{1.9}$$

The direction of this field must be at right angles everywhere both to each element of the wire and to r. Since we are representing the field by lines of force, the same must hold true for these. A little thought makes it clear that only one pattern will satisfy these requirements. The lines of force must form circles round the wire. Figure 1.9 shows what they look like if the wire is perpendicular to this page and the current is flowing away from you downwards into the page. (The vector product in the Biot–Savart equation corresponds to the field lines circulating in the same direction—clockwise—as a corkscrew rotating into the page.) Note that the spacing of the lines of force increases with distance from the wire because the field strength is falling off inversely with the distance.

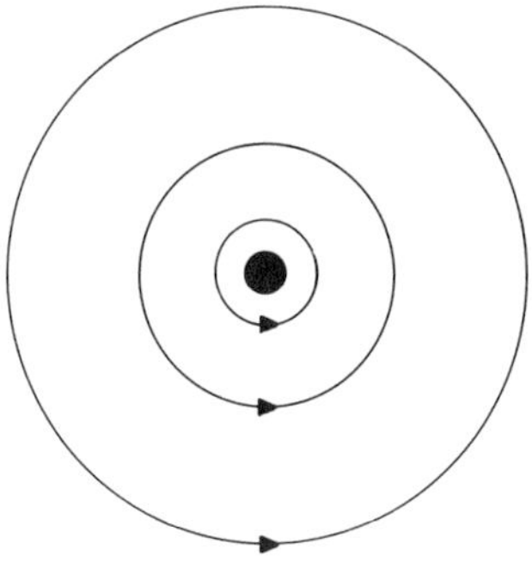

Figure 1.9

We now draw an imaginary closed surface round the wire (as shown in figure 1.10) and ask: what is the net flux of lines of force through this surface? The answer is obviously zero, since every exit point of a line through the surface has a corresponding entry point of the line somewhere else. We have defined net flux earlier as the divergence of $\boldsymbol{B}$. So we find that $\nabla\cdot\boldsymbol{B}=0$ for magnetic fields generated by currents, as for permanent magnets.

Since looking at the divergence—that is, the net flux outwards from the wire—leads to nothing new, we need to try a different approach. A glance at the pattern of lines of force suggests that we should, instead, consider the integral of $\boldsymbol{B}$ along a closed path which goes round the wire. That is to say, we evaluate $\int_s \boldsymbol{B}\cdot\mathrm{d}\boldsymbol{s}$, where $\mathrm{d}\boldsymbol{s}$ is an element of length of the path. The situation is illustrated in figure 1.11(*b*) where the broken line represents the path of integration. As can be seen

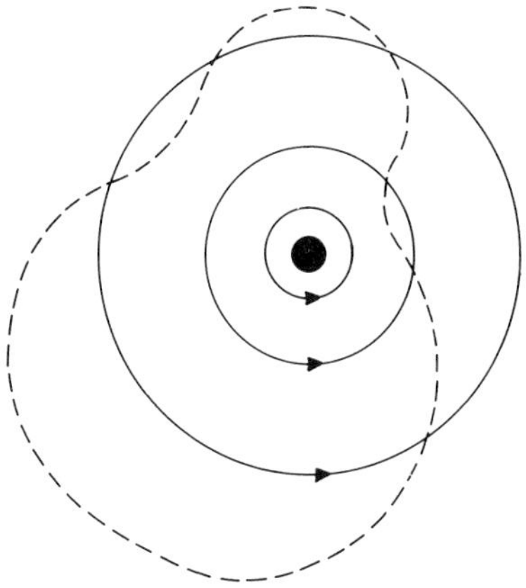

Figure 1.10

from figures 1.11(*a*) and (*b*), $\boldsymbol{B}\cdot\mathrm{d}\boldsymbol{s}=B\,\mathrm{d}s\cos\phi=B\,\mathrm{d}l=Bp\,\mathrm{d}\theta$. From equation (1.9) above, we can therefore write

$$\int_s \boldsymbol{B}\cdot\mathrm{d}\boldsymbol{s}=\int_0^{2\pi}\frac{\mu_0 i}{2\pi p}p\,\mathrm{d}\theta=\mu_0 i.$$

This is clearly the sort of equation we are looking for; it relates the magnetic field to the current producing the field. We can immediately extend its scope. It is similar to our previous equations in the sense that we can add the effects of a series of currents together. Indeed, we can generalise from a set of wires, each

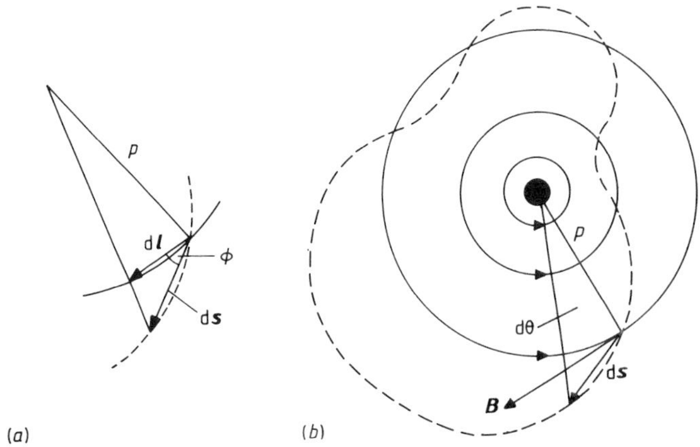

Figure 1.11

carrying a current, to a continuous distribution of current (in much the same way that, in our equation for $\boldsymbol{E}$, we moved from distinct electrical charges to a continuous distribution of charge ρ). In this case, we normally talk about a current density $\boldsymbol{j}$ per unit area. The current $i=\int_A \boldsymbol{j}\cdot \mathrm{d}\boldsymbol{A}$ (where $\mathrm{d}\boldsymbol{A}$ is an element of area; see figure 1.12). This allows us to rewrite our equation in the form

$$\int_s \boldsymbol{B}\cdot \mathrm{d}\boldsymbol{s}=\mu_0 \int_A \boldsymbol{j}\cdot \mathrm{d}\boldsymbol{A}.$$

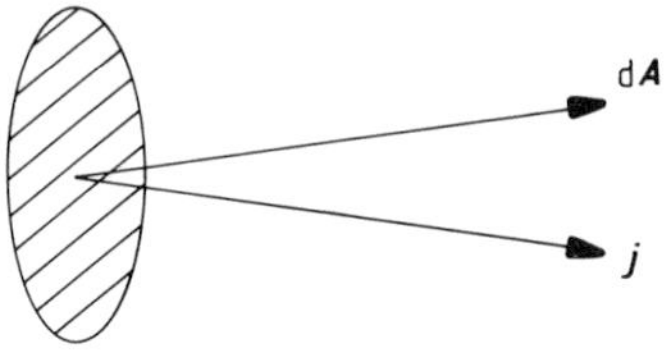

Figure 1.12

We can also change the left-hand side of this equation. The curl of a vector $\boldsymbol{B}$ is defined by

$$\text{curl } \boldsymbol{B}=\operatorname*{Lim}_{A\to 0}\left(\int_s \boldsymbol{B}\cdot \mathrm{d}\boldsymbol{s}\right)\frac{1}{A}$$

where, as before, $\mathrm{d}\boldsymbol{s}$ is an element of length of a path which here runs round the perimeter of an area A. It follows that

$$\int_s \boldsymbol{B}\cdot \mathrm{d}\boldsymbol{s}=\int_A \text{curl } \boldsymbol{B}\cdot \mathrm{d}\boldsymbol{A}. \tag{1.10}$$

This mathematical result is sometimes known as Stokes' theorem. We can therefore write

$$\int_A \text{curl } \boldsymbol{B}\cdot \mathrm{d}\boldsymbol{A}=\mu_0 \int_A \boldsymbol{j}\cdot \mathrm{d}\boldsymbol{A}.$$

Differentiating this equation, we have

$$\text{curl } \boldsymbol{B}=\mu_0 \boldsymbol{j}.$$

We can put this in a form corresponding to that used for the divergence of $\boldsymbol{B}$ and $\boldsymbol{E}$. We have seen that div $\boldsymbol{B}$ can be written as $\nabla\cdot\boldsymbol{B}$. Similarly, we can write curl $\boldsymbol{B}$ as $\nabla\wedge\boldsymbol{B}$. In cartesian coordinates:

$$\begin{aligned}\nabla\wedge\boldsymbol{B}&=\left(\boldsymbol{i}\frac{\partial}{\partial x}+\boldsymbol{j}\frac{\partial}{\partial y}+\boldsymbol{k}\frac{\partial}{\partial z}\right)\wedge(\boldsymbol{i}B_x+\boldsymbol{j}B_y+\boldsymbol{k}B_z)\\ &=\boldsymbol{i}\left(\frac{\partial B_z}{\partial y}-\frac{\partial B_y}{\partial z}\right)+\boldsymbol{j}\left(\frac{\partial B_x}{\partial z}-\frac{\partial B_z}{\partial x}\right)+\boldsymbol{k}\left(\frac{\partial B_y}{\partial x}-\frac{\partial B_x}{\partial y}\right).\end{aligned}$$

The equation

$$\nabla \wedge \boldsymbol{B} = \mu_0 \boldsymbol{j} \tag{1.11}$$

is known as Ampère's law, after the nineteenth century French scientist.

1.4 Varying Electric Field

We now consider magnetic and electric fields which vary with time. The problem with equation (1.11) is that, although the equation links $\boldsymbol{B}$ and $\boldsymbol{j}$, it contains no reference to changes with time. Indeed, we derived it purely in terms of a constant current. Let us see what this implies.

We can draw an imaginary closed surface round any part of an electrical circuit, and ask what is the net current flowing through this surface (see figure 1.13). The answer is obviously $\int_A \boldsymbol{j} \cdot \mathrm{d}\boldsymbol{A}$. If this is positive—a net outflow of current—there must clearly be an accompanying reduction of charge within the volume enclosed by our surface. We will take the charge density, as before, to be ρ. Then the total charge within the volume is $\int_V \rho \, \mathrm{d}V$ and the rate of loss of this charge from the volume is $(\partial/\partial t)\int_V \rho \, \mathrm{d}V$. Since this loss is producing the additional current, we can write

$$\int_A \boldsymbol{j} \cdot \mathrm{d}\boldsymbol{A} = -\frac{\partial}{\partial t} \int_V \rho \, \mathrm{d}V.$$

(The minus sign is introduced because an outward flow of current corresponds to a decrease in the amount of charge; see figure 1.13).

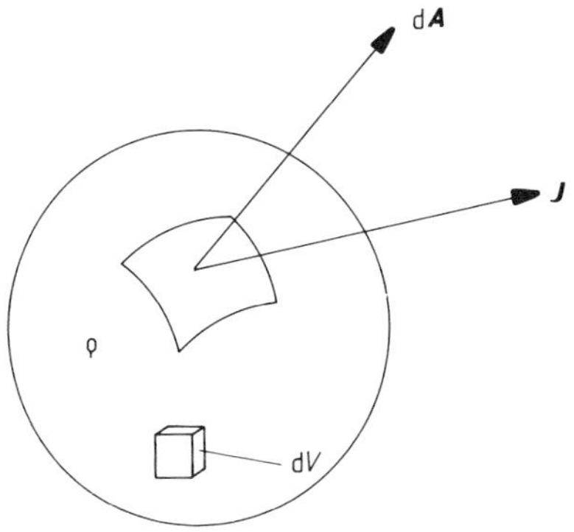

Figure 1.13

The left-hand side of this equation has a familiar form. Referring back to equation (1.4) for the electric field, we see that it can, by analogy, be related to the definition of the divergence of $\boldsymbol{j}$. In fact,

$$\int_V \operatorname{div} \boldsymbol{j}\, \mathrm{d}V = \int_A \boldsymbol{j} \cdot \mathrm{d}\boldsymbol{A}.$$

Our equation therefore becomes

$$\int_V \nabla \cdot \boldsymbol{j}\, \mathrm{d}V = -\frac{\partial}{\partial t} \int_V \rho\, \mathrm{d}V.$$

Differentiating both sides with respect to the volume finally gives us

$$\nabla \cdot \boldsymbol{j} = -\frac{\partial \rho}{\partial t}. \qquad (1.12)$$

This represents the equation of continuity for electric charge (so called because it implies that charges can be neither created, nor destroyed).

We now return to equation (1.11). If we take the divergence of both sides of this equation (that is multiply both sides by the operator $\nabla \cdot$), we have

$$\nabla \cdot (\nabla \wedge \boldsymbol{B}) = \mu_0 \nabla \cdot \boldsymbol{j}.$$

The left-hand side has the form $\boldsymbol{A} \cdot (\boldsymbol{A} \wedge \boldsymbol{B})$, where $\boldsymbol{A}$ and $\boldsymbol{B}$ are any two vectors. The term $\boldsymbol{A} \wedge \boldsymbol{B}$ is at right angles to both $\boldsymbol{A}$ and $\boldsymbol{B}$. The scalar product of $\boldsymbol{A}$ with $\boldsymbol{A} \wedge \boldsymbol{B}$ must therefore be zero, since the two vectors are at right angles (see figure 1.14). It follows that $\nabla \cdot (\nabla \wedge \boldsymbol{B}) = 0$. In physical terms, this corresponds to the fact that curl $\boldsymbol{B}$ follows the lines of force, for which, as we have seen, the divergence is zero.

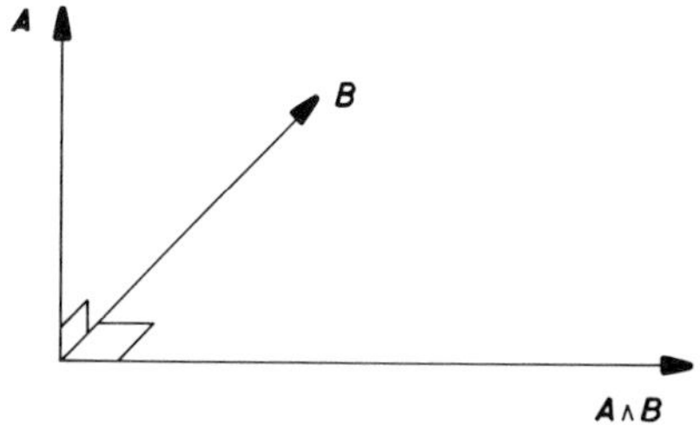

Figure 1.14

Consequently Ampère's equation (1.11) implies $\nabla \cdot \boldsymbol{j} = 0$. Comparing this result with the equation of continuity (1.12), we see that Ampère's equation implies $\partial \rho / \partial t = 0$. In other words, it only holds for steady currents, where the rate of flow of charge is constant. To obtain the full equation for $\boldsymbol{B}$, we must allow for cases where $\partial \rho / \partial t \neq 0$.

We can try to extend equation (1.11) by adding a term to the right-hand side. This term must evidently have the same dimensions as current density, if the equation is to balance, so we can rewrite it as

$$\nabla \wedge \boldsymbol{B}=\mu_0(\boldsymbol{j}+\boldsymbol{j}_\mathrm{D}) \tag{1.13}$$

where the new term we have introduced, $\boldsymbol{j}_\mathrm{D}$, is called the 'displacement' current. If we now take the divergence of both sides, we have

$$\nabla\cdot(\nabla \wedge \boldsymbol{B})=0=\mu_0(\nabla\cdot\boldsymbol{j}+\nabla\cdot\boldsymbol{j}_\mathrm{D}).$$

This means that

$$\nabla\cdot\boldsymbol{j}_\mathrm{D}=-\nabla\cdot\boldsymbol{j}.$$

However, we have seen that $\nabla\cdot\boldsymbol{j}=-\partial\rho/\partial t$, so it follows that $\nabla\cdot\boldsymbol{j}_\mathrm{D}=\partial\rho/\partial t$.

We have previously derived a relationship between ρ and $\boldsymbol{E}$ in equation (1.5), namely that

$$\nabla\cdot\boldsymbol{E}=\rho/\varepsilon_0.$$

Differentiating with respect to time, and using the relationship between $\nabla\cdot\boldsymbol{j}_\mathrm{D}$ and $\partial\rho/\partial t$ leads to

$$\frac{\partial}{\partial t}(\nabla\cdot\boldsymbol{E})=\nabla\cdot\frac{\partial\boldsymbol{E}}{\partial t}=\frac{1}{\varepsilon_0}\frac{\partial\rho}{\partial t}=\frac{1}{\varepsilon_0}\nabla\cdot\boldsymbol{j}_\mathrm{D}.$$

We can rewrite this in the form

$$\nabla\cdot\left(\frac{\partial\boldsymbol{E}}{\partial t}-\boldsymbol{j}_\mathrm{D}/\varepsilon_0\right)=0.$$

It follows that the equation can be reduced to $\partial\boldsymbol{E}/\partial t=\boldsymbol{j}_\mathrm{D}/\varepsilon_0$. Substituting back into our original equation (1.13) gives us finally

$$\nabla \wedge \boldsymbol{B}=\mu_0\boldsymbol{j}+\varepsilon_0\mu_0\frac{\partial\boldsymbol{E}}{\partial t}.$$

It is customary to write $1/\varepsilon_0\mu_0=c^2$, so that this becomes

$$c^2\nabla \wedge \boldsymbol{B}=\boldsymbol{j}/\varepsilon_0+\frac{\partial\boldsymbol{E}}{\partial t} \tag{1.14}$$

This is the third of Maxwell's equations.

1.5 Varying Magnetic Field

We come to the last of Maxwell's equations. We have just dealt with the magnetic field produced by changing electrical effects. The remaining equation must obviously relate a changing magnetic field to the electrical effects it produces. Over a century and a half ago, the English scientist Faraday found that when the magnetic flux through a conducting circuit changed, a current flowed in the circuit. This 'induced' current flows in such a way as to oppose the original change. (This is sometimes called Lenz's law, after the Russian scientist who suggested it.) The induced current produces its own magnetic field and the direction of flow of the current is such that this secondary magnetic field acts in the opposite direction to the original magnetic field.

The induced current is actually a secondary effect: its value depends on the resistance of the circuit. What the change in magnetic flux leads to in the first instance is an electromotive force $\mathscr{E}$ which then produces the observed current. We can write

$$\mathscr{E} = -\mathrm{d}N/\mathrm{d}t \tag{1.15}$$

where the magnetic flux is represented by the number of lines of force N and the minus sign reflects the opposition to change.

A simple example of this is provided by a bar magnet which is moved towards a wire ring. As it moves, the number of lines of force threading their way through the ring will change. This leads to an electromotive force, and so to a current in the ring, which produces a magnetic field (shown in figure 1.15 by dotted circles). The amount of induced current (and, therefore, the strength of the opposing magnetic field) depends on how rapidly the magnet is moved towards the circuit; the faster the motion, the more lines of force cut the circuit per unit time.

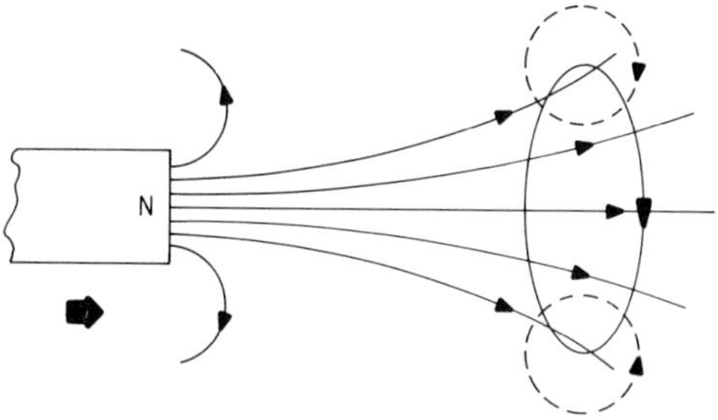

Figure 1.15

We need to convert equation (1.15) into the same form as the other Maxwell equations. We begin with the left-hand side of this equation by noting that the induced electromotive force corresponds to the electric field taken all round the circuit. This follows simply from our definition of the electric field at the beginning of this chapter as the force acting on electric charges. Here, the charges are normally contained within the conducting elements of the circuit, and the force is integrated over each element. We label each element d$\boldsymbol{s}$, and the electric field $\boldsymbol{E}$, as before, and can then write

$$\mathscr{E} = \int_s \boldsymbol{E} \cdot \mathrm{d}\boldsymbol{s}. \tag{1.16}$$

Reference back to our derivation of Maxwell's third equation (1.10) shows that the right-hand side of equation (1.16) is related to the curl of the vector. In fact, we can rewrite it as

$$\int_s \boldsymbol{E}\cdot \mathrm{d}\boldsymbol{s} = \int_A \mathrm{curl}\ \boldsymbol{E}\cdot \mathrm{d}\boldsymbol{A}$$

where $\mathrm{d}\boldsymbol{A}$ is an element of the area enclosed by the circuit.

We can also reformulate the right-hand side of equation (1.15). We have seen previously in equation (1.6) that the number of lines of force N can be related to the magnetic field $\boldsymbol{B}$ by the equation $N=\int_A \boldsymbol{B}\cdot \mathrm{d}\boldsymbol{A}$. Differentiating both sides, this gives

$$\frac{\mathrm{d}N}{\mathrm{d}t} = \int_A \frac{\partial \boldsymbol{B}}{\partial t}\cdot \mathrm{d}\boldsymbol{A}.$$

Note that there is a partial derivative on the right-hand side because the magnetic field varies with position as well as time.

Combining the rewritten forms for both sides of equation (1.15), we have

$$\int_A \mathrm{curl}\ \boldsymbol{E}\cdot \mathrm{d}\boldsymbol{A} = -\int_A \frac{\partial \boldsymbol{B}}{\partial t}\cdot \mathrm{d}\boldsymbol{A}.$$

Differentiating both sides, as before,

$$\mathrm{curl}\ \boldsymbol{E} = -\frac{\partial \boldsymbol{B}}{\partial t}$$

or, with our usual notation,

$$\nabla \wedge \boldsymbol{E} = -\frac{\partial \boldsymbol{B}}{\partial t} \qquad (1.17)$$

This is Maxwell's fourth equation.

1.6 The Lorentz Force

We began our examination of Maxwell's equations in §1.2 by looking at the force exerted by an electric charge. We have not yet considered the forces exerted by the magnetic field, so we next turn to these.

In §1.3 we discussed a current i flowing along a wire. Let us now suppose that this wire is immersed in a magnetic field $\boldsymbol{B}$. It is found experimentally that a force $\mathrm{d}\boldsymbol{F}$ is exerted by the field on a short length of the wire $\mathrm{d}\boldsymbol{s}$ such that (see figure 1.16)

$$\mathrm{d}\boldsymbol{F} = i(\mathrm{d}\boldsymbol{s} \wedge \boldsymbol{B}). \qquad (1.18)$$

Note that the force is at right angles both to the wire and to the magnetic field (since we are dealing with the vector product). This should be contrasted with the electric force, which acts in the direction of the field (see §1.1).

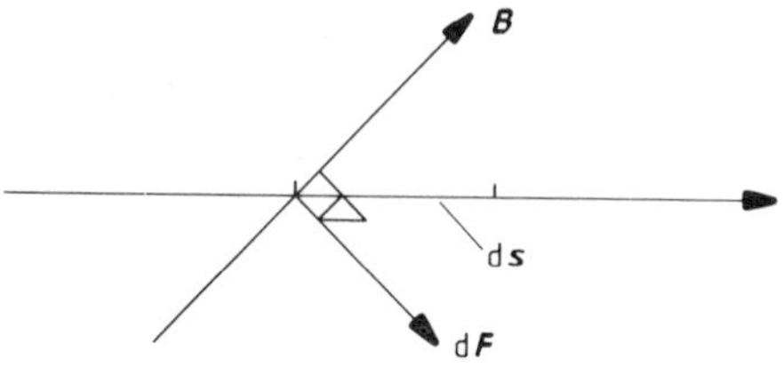

Figure 1.16

We can compare equation (1.18) with the Biot–Savart equation (1.8), which concerns the magnetic field produced by a current flowing in a wire. We see that both equations involve the term $i\mathrm{d}\boldsymbol{s}$. Suppose we think of the current in terms of an electric charge q flowing along the wire. We can obviously put $i=\mathrm{d}q/\mathrm{d}t$. If the charge is moving with a velocity $\boldsymbol{v}$ then $\mathrm{d}\boldsymbol{s}=\boldsymbol{v}\mathrm{d}t$. Hence, $i\mathrm{d}\boldsymbol{s}=(\mathrm{d}q/\mathrm{d}t)\boldsymbol{v}\mathrm{d}t=\mathrm{d}q\boldsymbol{v}$. We can therefore rewrite equation (1.8) in the form

$$\mathrm{d}\boldsymbol{B}=\frac{\mu_0}{4\pi r^3}(\mathrm{d}q\boldsymbol{v}\wedge\boldsymbol{r})$$

or, integrating over the charge,

$$\boldsymbol{B}=\frac{\mu_0 q}{4\pi r^3}(\boldsymbol{v}\wedge\boldsymbol{r}). \tag{1.19}$$

Similarly, we can rewrite equation (1.18) as

$$\mathrm{d}\boldsymbol{F}=\mathrm{d}q(\boldsymbol{v}\wedge\boldsymbol{B})$$

or, integrating again,

$$\boldsymbol{F}=q(\boldsymbol{v}\wedge\boldsymbol{B}). \tag{1.20}$$

If we now insert equation (1.19) into equation (1.20), we obtain the following expression for the magnetic force

$$\boldsymbol{F}_{\mathrm{mag}}=\frac{\mu_0 q^2}{4\pi r^3}[\boldsymbol{v}\wedge(\boldsymbol{v}\wedge\boldsymbol{r})].$$

Now, compare this with the electric force given by equation (1.1),

$$\boldsymbol{F}_{\mathrm{elect}}=\frac{q^2}{4\pi\varepsilon_0 r^3}\boldsymbol{r}.$$

We ignore for the moment the vectorial nature of these two equations and consider only their magnitudes. Dividing one by the other, we have

$$\frac{F_{\mathrm{mag}}}{F_{\mathrm{elect}}}=\varepsilon_0\mu_0 v^2=\frac{v^2}{c^2} \tag{1.21}$$

where we have used the notation for $\varepsilon_0\mu_0$ introduced in equation (1.14). We see that the constant c must have the dimensions of a velocity and its magnitude is the velocity of light in a vacuum.

In many electromagnetic situations, electric and magnetic forces will be present simultaneously. In §1.1, we defined the electric field by

$$\boldsymbol{F}_{\mathrm{elect}}=q\boldsymbol{E}.$$

Adding this to equation (1.20) gives

$$\boldsymbol{F}_{\mathrm{total}}=\boldsymbol{F}_{\mathrm{elect}}+\boldsymbol{F}_{\mathrm{mag}}=q(\boldsymbol{E}+\boldsymbol{v}\wedge\boldsymbol{B}) \tag{1.22}$$

$\boldsymbol{F}_{\mathrm{total}}$ is known as the Lorentz force, after the Dutch scientist who worked on it at the end of the nineteenth century. Although we have derived this force in terms of a current flowing along a wire, it applies equally to an individual charged particle moving through electric and magnetic fields.

The second term in equation (1.22), $\boldsymbol{v}\wedge\boldsymbol{B}$, represents a kind of additional electric field which acts on any charged body in motion. It needs to be kept in mind whenever magnetic fields are present. As an example, we will now look at Ohm's law. This law, named after the nineteenth century German scientist who established it experimentally, is obviously concerned with current flow—that is, with charges in motion. It is usually written in the form

$$i=\mathscr{E}/R$$

where i is the current, $\mathscr{E}$ is the electromotive force and R is the resistance of the circuit.

We can reformulate Ohm's law in a more general way to parallel the form in which we have written Maxwell's equations. Consider a small element within a wire of length $\mathrm{d}s$ and cross sectional area $\mathrm{d}A$ as shown in figure 1.17. Instead of considering the resistance of the entire wire, we can consider the resistivity ρ of the wire. The two are related, assuming the wire to be homogeneous, by

$$R=\rho\,\mathrm{d}s/\mathrm{d}A.$$

It is customary to replace the resistivity of the material by its inverse, the conductivity $\sigma=1/\rho$. We can then write

$$R=\frac{1}{\sigma}\frac{\mathrm{d}s}{\mathrm{d}A}.$$

The current flowing through the small element of wire can be written in terms of the current density

$$i=j\,\mathrm{d}A.$$

Finally, the electromotive force—as we saw in equation (1.16)—can also be rewritten as

$$\mathscr{E} = E\,ds.$$

Substituting all these into the original form of Ohm's law leads to the reformulated version

$$j = \sigma E.$$

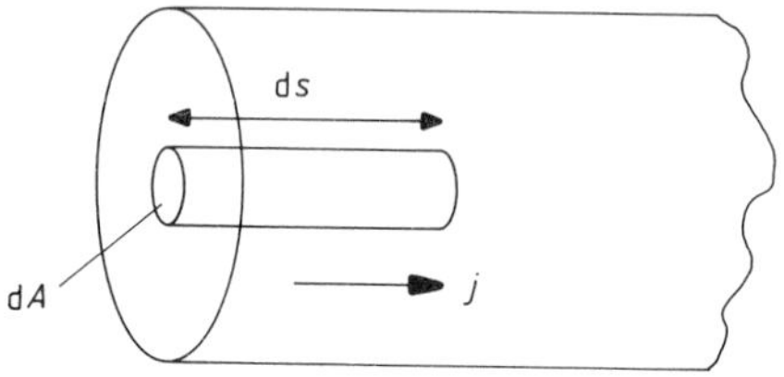

Figure 1.17

A comparison of this with equation (1.22) indicates that Ohm's law needs to be modified in the presence of a magnetic field. The modified form (introducing vectorial notation again) must obviously be

$$\boldsymbol{j} = \sigma(\boldsymbol{E} + \boldsymbol{v} \wedge \boldsymbol{B}). \tag{1.23}$$

1.7 Electromagnetic Units

A brief word now about the units used in electromagnetic calculations. They should obviously be consistent across the entire range of calculations. Historically, this has not always been true, because electric fields and magnetic fields have often been treated separately. However, the units used in this book are consistent across the board. It is also convenient to employ the units which are actually used in laboratory measurements—the ampere, the volt and so on. The constants ε_0 and μ_0 have been introduced into Maxwell's equations to make this possible. In numerical terms,

$$\varepsilon_0 = \frac{1}{4\pi}\frac{10^{-9}}{9}\,\mathrm{F\,m^{-1}} \qquad \text{and} \qquad \mu_0 = 4\pi \times 10^{-7}\,\mathrm{H\,m^{-1}}.$$

The factor 4π in each of these numerical values arises from our basic definitions. For example, we have written the constant in equation (1.1) as $1/4\pi\varepsilon_0$, rather than $1/\varepsilon_0$. The reason for introducing this factor can be seen in our derivation of Maxwell's equations. We have repeatedly talked about 'drawing an imaginary surface' round a region of space. This implies introducing a solid angle of 4π: thus, drawing a sphere of unit radius round a point corresponds to introducing a surface of area 4π. Putting 4π into the basic equations therefore allows us to eliminate it from subsequent equations.

The units we are using in this book, which take all these points into account, are usually called SI units. (SI stands for 'International System'; the letters are the other way round because they are abbreviated from the French equivalent 'Système International'.) They are sometimes referred to as 'rationalised mks' units. 'Rationalised' simply means that the factor 4π has been introduced into the basic equations. The 'mks' stands for 'metre, kilogram, second'. These are the basic practical units of measurement in mechanics, corresponding to the ampere etc in electromagnetism. The term is therefore used as shorthand for a system using practical units.

1.8 Summary

Finally we gather together the four Maxwell equations and place them in a box. The Lorentz force law and the equation of continuity have been added below, since these two equations are often required in applications of Maxwell's equations.

$$\nabla \cdot \boldsymbol{E} = \rho/\varepsilon_0 \tag{1.5}$$

$$\nabla \cdot \boldsymbol{B} = 0 \tag{1.7}$$

$$c^2 \nabla \wedge \boldsymbol{B} = \boldsymbol{j}/\varepsilon_0 + \partial \boldsymbol{E}/\partial t \tag{1.14}$$

$$\nabla \wedge \boldsymbol{E} = -\partial \boldsymbol{B}/\partial t \tag{1.17}$$

$$\boldsymbol{F} = q(\boldsymbol{E} + \boldsymbol{v} \wedge \boldsymbol{B}) \tag{1.22}$$

$$\nabla \cdot \boldsymbol{j} = -\partial \rho/\partial t \tag{1.12}$$

Maxwell's Equations—Applications

2

2.1 Introduction

In the first chapter we have obtained a set of four partial differential equations. These equations, together with the Lorentz force law, constitute a complete description of electromagnetism—the Maxwell theory. We should qualify the last sentence by stressing that Maxwell's theory is a classical theory: it applies so long as quantum effects are insignificant. For example, we must ignore the fact that light sometimes behaves as though it is composed of photons. Notwithstanding this limitation of the theory, its range of applications is enormously large. It embraces all macroscopic electromagnetic phenomena, such as the force laws between charges and between currents, the workings of dynamos and motors and the propagation of radio waves through material media and through a vacuum. In the first chapter we were concerned with making Maxwell's equations plausible, by showing how they originated from experimental laws of electricity and magnetism discovered in the nineteenth century. In this chapter we will start by knocking away the scaffolding we used to obtain the equations. We will now accept the equations as the basis of electromagnetic theory, and adopt them as the starting point for our investigations into electromagnetism.

We start by considering the special case of static fields. A static field is one which does not change with time; therefore the $\boldsymbol{E}$ and $\boldsymbol{B}$ fields satisfy $\partial \boldsymbol{E}/\partial t = \partial \boldsymbol{B}/\partial t = 0$. Inserting these conditions into Maxwell's equations, which are quoted in the box at the end of Chapter 1, we get

$$\nabla \cdot \boldsymbol{E} = \rho/\varepsilon_0 \tag{2.1}$$

$$\nabla \wedge \boldsymbol{E} = 0 \tag{2.2}$$

$$\nabla \cdot \boldsymbol{B} = 0 \tag{2.3}$$

$$c^2 \nabla \wedge \boldsymbol{B} = \boldsymbol{j}/\varepsilon_0. \tag{2.4}$$

The left-hand sides of the above equations are independent of time, since the operator ∇ contains only derivatives with respect to the spatial coordinates. Consequently, the right-hand sides must also be time independent, so ρ and $\boldsymbol{j}$ are static, like the fields. Another consequence of the disappearance of the $\partial \boldsymbol{E}/\partial t$ and

$\partial \boldsymbol{B}/\partial t$ terms from the right-hand sides of equations (2.2) and (2.4) is that the $\boldsymbol{E}$ and $\boldsymbol{B}$ fields no longer depend on each other. Thus, in the static case, Maxwell's equations separate into two independent pairs of equations. The first pair of equations governs the electric fields produced by static arrays of charges: they are the equations of electrostatics. The second pair are the equations of magnetostatics and govern the magnetic fields produced by steady currents.

2.2 Electrostatics

We begin our investigation into the predictions of Maxwell's equations by considering the special case of electrostatic fields. We have just seen that in this case Maxwell's equations reduce to the pair of equations

$$\nabla \cdot \boldsymbol{E} = \rho/\varepsilon_0 \tag{2.1}$$

$$\nabla \wedge \boldsymbol{E} = 0. \tag{2.2}$$

If we are given the charge density ρ as a function of position, then these two equations can be solved by one means or another for the field $\boldsymbol{E}$. For most distributions of charge, this task will present mathematical difficulties and a great deal of skill and hard work may be needed to obtain a solution. However, for charge distributions which possess a high degree of symmetry the field can be obtained easily, as we will now show. We need Gauss's law (equation (2.1)) in integral form, which is

$$\int_A \boldsymbol{E} \cdot \mathrm{d}\boldsymbol{A} = \frac{1}{\varepsilon_0} \int_V \rho \, \mathrm{d}V \tag{2.5}$$

(see equation (1.3)). Now, by choosing a closed surface A which has the same symmetry as the charge distribution, we can evaluate the integrals in equation (2.5) without difficulty. As an illustration, let us obtain the field due to a long line of charge having uniform charge per unit length σ. As the charge distribution has cylindrical symmetry we choose a cylindrical surface of radius r and unit length coaxial with the line of charge and two end discs of radius r to close the surface. This imaginary surface over which the integration is to be performed is called a *gaussian surface.* The arrangement is shown in figure 2.1. From symmetry considerations it is clear that the magnitude of the field is constant over the cylindrical part of the gaussian surface provided that the ends of the line of charge are well beyond the ends of the cylinder. But what about the direction of the field? Figure 2.2 shows two field configurations which have cylindrical symmetry. The equation $\nabla \wedge \boldsymbol{E} = 0$ will enable us to decide between these two possibilities. In integral form (see equation (1.10)) this equation is

$$\int_s \boldsymbol{E} \cdot \mathrm{d}\boldsymbol{s} = 0. \tag{2.6}$$

Equation (2.6) tells us that if we integrate $\boldsymbol{E}$ round any closed path in the electrostatic field the result must be zero. Clearly if the integration is carried out around the circular path depicted in figure 2.2(*b*) the result is non-zero, so this configuration is not allowed. Integrating around the same path in figure 2.2(*a*) yields a value of zero; in fact integration round any closed path in this field yields zero. Thus the field lines are radial to the line of charge. We are now in a position to carry out the integration of equation (2.5). The left-hand side must be evaluated over the curved cylindrical surface and over the two end discs. As the field lines are parallel to the flat end discs there is no flux of $\boldsymbol{E}$ across these two surfaces, so their contribution to the integral is zero. This leaves an integral over the curved cylindrical surface. $\boldsymbol{E}$ is everywhere constant and perpendicular to this surface, so

$$\int_A \boldsymbol{E}\cdot \mathrm{d}\boldsymbol{A} = E\int_A \mathrm{d}A = 2\pi r E$$

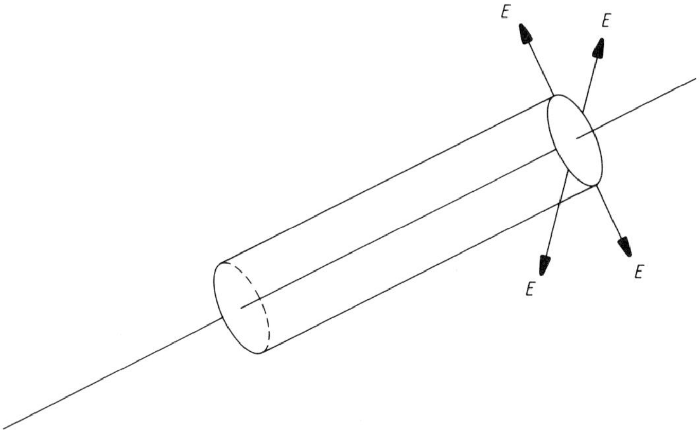

Figure 2.1 The line of charge and the cylindrical gaussian surface over which the integral in equation (2.5) is evaluated. The surface consists of the curved part and two end discs. It has unit length and radius r.

where E is the magnitude of the electrostatic field. The integral on the right-hand side is even easier to evaluate; it is the total charge within the volume enclosed by the gaussian surface, which is just the charge on unit length of the line, i.e. σ. Thus we find

$$E = \frac{\sigma}{2\pi\varepsilon_0 r}. \tag{2.7}$$

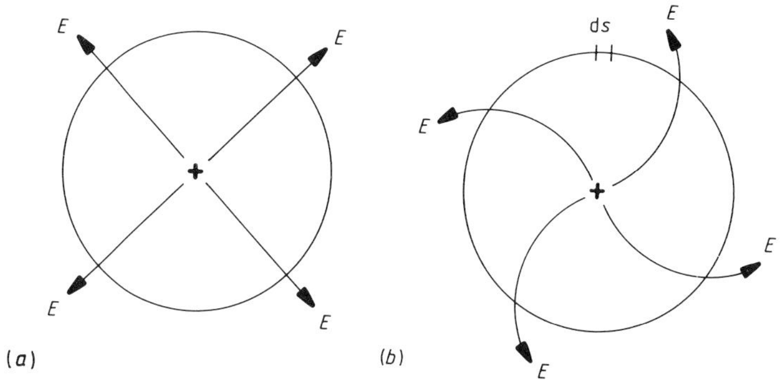

Figure 2.2 Two proposed configurations for the field lines produced by a long line of charge, viewed along the line. Only (*a*) is consistent with the requirement that the electrostatic field must have vanishing curl.

Unfortunately there are only a limited number of problems where the symmetry enables us to integrate equation (2.5) so effortlessly. Nevertheless, these simple cases do illustrate the process of determining the form of the electrostatic field from Maxwell's equations.

2.3 Magnetostatics

We will now consider situations where there are steady currents flowing. These give rise to steady $\boldsymbol{B}$ fields whose governing equations are, as we have seen in §2.2,

$$\nabla \cdot \boldsymbol{B} = 0 \tag{2.3}$$

$$c^2 \nabla \wedge \boldsymbol{B} = \boldsymbol{j}/\varepsilon_0. \tag{2.4}$$

The differences between these equations and equations (2.1) and (2.2) arise from two important differences between the magnetostatic and the electrostatic field. The first difference is that there are no magnetic charges (or monopoles as they are usually known) on which lines of $\boldsymbol{B}$ can start or finish, so lines of $\boldsymbol{B}$ form closed loops and therefore the next flux out of any closed volume in the magnetostatic field is always zero. Equation (2.3) expresses this fact. The second difference is that the sources of the $\boldsymbol{B}$ field are currents which are vector quantities, as opposed to the sources of the $\boldsymbol{E}$ field which are scalar quantities. Thus the vector character of the two fields is different. To see this, compare the Biot–Savart law (equation (1.8)) with Coulomb's law (equation (1.2)).

If $\boldsymbol{j}$ is given as a function of position, the equations of magnetcstatics can in principle be solved for the field $\boldsymbol{B}$. In practice, this is generally difficult. If there is some symmetry present in the current distribution then the task is made easier.

For example, in Chapter 1 the field due to a long straight wire carrying a steady current was found by evaluating the integral in equation (1.8).

An interesting field distribution is that produced by a current flowing in a closed circular loop of wire. We can imagine that the wire is a superconductor, so that once started the current will flow forever (a superconductor is a material whose electrical resistance is zero). Bearing in mind the form of the field around a long straight wire we can, with a little thought, convince ourselves that the pattern of field lines must be of the form shown in figure 2.3. Close to the wire the lines of $\boldsymbol{B}$ will be very nearly circular but further away from the wire and inside the loop we can think of the lines trying to repel each other and thus the circles become flattened as shown. Such a field is called a dipole field: away from the loop it has the same appearance as the field of a short bar magnet situated at the centre of the loop and oriented perpendicular to it. As Maxwell's equations tell us that steady currents are the only source of magnetostatic fields, we can deduce that there must be permanent atomic current loops giving rise to the bar magnet's field.

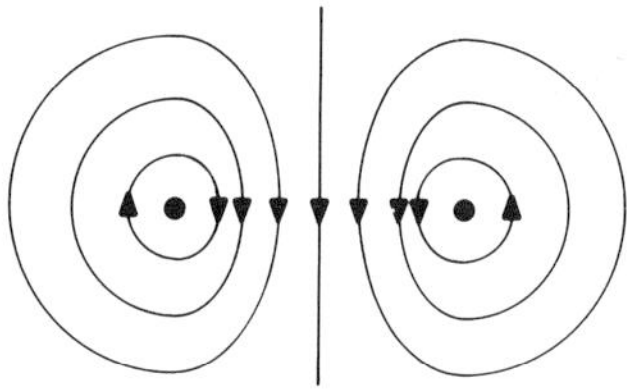

Figure 2.3

The field $\boldsymbol{B}$ due to the loop may be evaluated at a general point by means of the Biot–Savart law. This law (equation (1.8)) states that

$$\mathrm{d}\boldsymbol{B}=\frac{\mu_0 i}{4\pi r^3}(\mathrm{d}\boldsymbol{s}\wedge\boldsymbol{r})$$

where $\mathrm{d}\boldsymbol{B}$ is the increment of field produced by a current element of length $\mathrm{d}\boldsymbol{s}$ a distance r away from the field point. The total field $\boldsymbol{B}$ due to the whole loop can be obtained by integrating the contributions from all elements of the loop. For an arbitrary field point this integration is difficult to perform, but we can obtain $\boldsymbol{B}$ on the axis of the loop simply, as follows. Let the loop have radius a and consider the field at a point P distance d above the centre of the loop (see figure 2.4). Then the contribution from the current element $\mathrm{d}\boldsymbol{s}$ is

$$\mathrm{d}\boldsymbol{B}=\frac{\mu_0 i\,\mathrm{d}s\boldsymbol{n}}{4\pi r^2}$$

where $\boldsymbol{r}$ is perpendicular to $\mathrm{d}\boldsymbol{s}$ so $\boldsymbol{r}\wedge\mathrm{d}\boldsymbol{s}=r\,\mathrm{d}s\boldsymbol{n}$ and $\boldsymbol{n}$ is a unit vector perpendicular to $\boldsymbol{r}$ and d$\boldsymbol{s}$. Note that the contributions d$\boldsymbol{B}$ from different elements are in different directions, but from symmetry it follows that all components of $\boldsymbol{B}$ not parallel to the axis cancel. The resultant field is therefore obtained by summing the axial components of d$\boldsymbol{B}$ due to all elements in the loop, and thus

$$|\boldsymbol{B}| = \frac{\mu_0 i}{4\pi} \frac{\cos\theta}{r^2} \int \mathrm{d}s = \frac{\mu_0 i a^2}{2(a^2 + d^2)^{3/2}}. \tag{2.8}$$

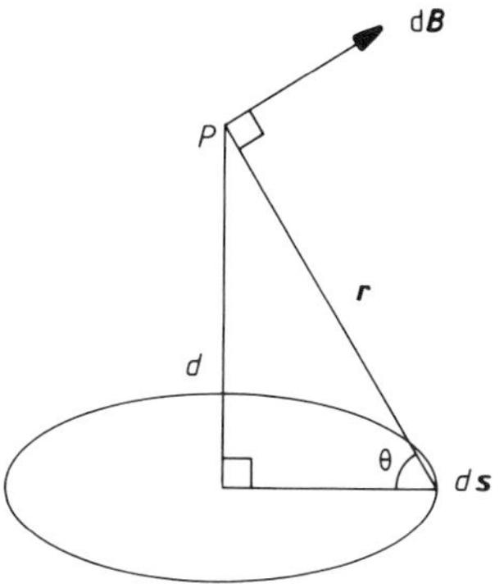

Figure 2.4

2.4 Electromagnetic Induction

The situations treated in the last two sections are highly idealised. In general we do not encounter perfectly stationary charges or steady currents. In the final two sections of this chapter we will show that in the more general case of time-varying fields Maxwell's equations predict that changes in the fields propagate at the speed of light. However, before doing that we consider in this section some effects associated with electromagnetic induction in which the finite speed of propagation of time-varying fields can be ignored. We may make this approximation when the travel time of light across the region of interest is small compared with the time for the field to change by an appreciable amount.

Let us suppose that we have a cylindrical volume within which there is a uniform $\boldsymbol{B}$ field. By uniform we mean that $\boldsymbol{B}$ does not vary with *position*. Such a field could be produced in the interior of a long solenoid for example. Let us also suppose, for the moment, that the $\boldsymbol{B}$ field is changing linearly with time, then $\partial \boldsymbol{B}/\partial t$ is constant. Now equation (1.17) states that

$$\nabla \wedge \boldsymbol{E} = -\frac{\partial \boldsymbol{B}}{\partial t} \tag{1.17}$$

so the time-varying $\boldsymbol{B}$ field gives rise to an $\boldsymbol{E}$ field. Moreover, in our case, $\nabla \wedge \boldsymbol{E}$ is independent of time so the $\boldsymbol{E}$ field must be independent of time as the ∇ operator consists of derivatives with respect to the position coordinates only. Therefore we have a static electric field yet one that is not generated by charges. As there are no charges we have from equation (1.5) that $\nabla \cdot \boldsymbol{E} = 0$. Thus the $\boldsymbol{E}$ field generated by the changing $\boldsymbol{B}$ field has finite curl but vanishing divergence just like the magnetostatic $\boldsymbol{B}$ field (see equations (2.3) and (2.4)). Now in §1.3 we determined the $\boldsymbol{B}$ field around a long straight wire carrying a constant current, an arrangement with the same cylindrical symmetry that we have here, and showed that the lines of $\boldsymbol{B}$ form closed loops around the wire as shown in figure 1.4. So in the present case, bearing in mind the similar symmetry, we deduce that the electric field lines will form closed loops concentric with the axis of the field region and perpendicular to it as shown in figure 2.5. However the variation of the magnitude of $\boldsymbol{E}$ with distance from the axis will not be the same as that of a $\boldsymbol{B}$ field produced by a steady current confined to a wire.

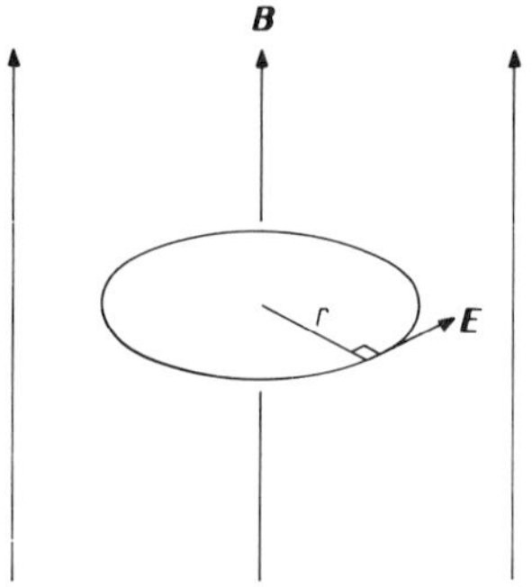

Figure 2.5

We can obtain the $\boldsymbol{E}$ field as follows. Combining equations (1.15) and (1.16) gives

$$\int_s \boldsymbol{E} \cdot \mathrm{d}\boldsymbol{s} = -\frac{\mathrm{d}N}{\mathrm{d}t}$$

where N is the flux of $\boldsymbol{B}$ through the loop s.

Choosing s to be the circular path of radius r centred on the axis of the cylindrical region of field, as shown in figure (2.5), we get for the right-hand side using the definition of flux

$$\frac{\mathrm{d}N}{\mathrm{d}t} = \pi r^2 \frac{\mathrm{d}B}{\mathrm{d}t}.$$

Now $\boldsymbol{E}$ is tangential to the path of integration and from symmetry has the same value at all points on the path so the left-hand side is

$$\int_s \boldsymbol{E} \cdot \mathrm{d}\boldsymbol{s} = E 2\pi r.$$

Thus

$$E 2\pi r = -\pi r^2 \frac{\mathrm{d}B}{\mathrm{d}t}.$$

In order for $\mathrm{d}B/\mathrm{d}t$ to be uniform E must be a linear function of r, thus

$$\boldsymbol{E} = E_0 r \hat{\boldsymbol{s}}$$

where $\hat{\boldsymbol{s}}$ is a unit vector tangential to the circle of radius r and

$$E_0 = \frac{1}{2}\left|\frac{\partial B}{\partial t}\right|.$$

If we now relax our restriction that $\partial \boldsymbol{B}/\partial t$ is constant; the spatial form of $\boldsymbol{E}$ will not change but the magnitude of E will become time dependent, so we get for the E field produced by a time-varying uniform $\boldsymbol{B}$ field

$$\boldsymbol{E} = E_0(t) r \hat{\boldsymbol{s}}.$$

Consider now a practical effect of this induced $\boldsymbol{E}$ field. When a cylindrical metal ring is placed in the field region with its axis parallel to the $\boldsymbol{B}$ field's axis, the electric field causes a current to flow around the ring and this current gives rise to its own magnetic field. Lenz's law tells us that the ring's magnetic field will be in the opposite direction to the original $\boldsymbol{B}$ field (see figures 1.15 and 2.3). Now the ring with its magnetic field will behave like a magnet. We can think of it as having positive poles on one face and an equal number of negative poles on the other face. However, because the external $\boldsymbol{B}$ field is uniform the translational force it exerts on the positive poles will be equal and opposite to the translational force it exerts on the negative poles and the ring will not move. If however the metal ring were to be moved to the end of the solenoid it would experience a non-uniform $\boldsymbol{B}$ field similar to that depicted in figure 1.15. In this situation there is a net translational force on the ring because the field strengths at the two ends of the ring are not the same. If the solenoid stands vertically in the earth's gravitational field and a sinusoidal current is passed through its coils the ring can be made to hover. This is the principle of magnetic levitation.

2.5 Electromagnetic Waves in Empty Space

We are now going to consider the more general situation in which the fields are functions of time; under these circumstances the $\boldsymbol{E}$ and $\boldsymbol{B}$ fields are not independent of each other, and all four Maxwell equations are needed. Our aim is to show that time-varying fields can propagate as waves through empty space. Putting the conditions for empty space, namely $\rho=0$ and $\boldsymbol{j}=0$, into the full Maxwell equations (1.5), (1.7), (1.14) and (1.17) gives us the four equations

$$\nabla \cdot \boldsymbol{E}=0 \tag{2.9}$$

$$\nabla \cdot \boldsymbol{B}=0 \tag{2.10}$$

$$\nabla \wedge \boldsymbol{B}=\frac{1}{c^2}\frac{\partial \boldsymbol{E}}{\partial t} \tag{2.11}$$

$$\nabla \wedge \boldsymbol{E}=-\frac{\partial \boldsymbol{B}}{\partial t}. \tag{2.12}$$

We notice that there are two divergence equations (2.9) and (2.10) which tell us that the flux of $\boldsymbol{E}$ and of $\boldsymbol{B}$ out of any volume is zero, and two curl equations (2.11) and (2.12) which couple the $\boldsymbol{E}$ and $\boldsymbol{B}$ fields together. Let us now look at what the coupling of these latter two equations implies. Equation (2.12) tells us that if $\partial\boldsymbol{B}/\partial t$ is non-zero, then so is $\nabla \wedge \boldsymbol{E}$; but $\nabla \wedge \boldsymbol{E}$ can only be non-zero if $\boldsymbol{E}$ is a function of position. Furthermore, if $\partial\boldsymbol{B}/\partial t$ is changing in time, then $\nabla \wedge \boldsymbol{E}$ is a function of time also; this means that $\boldsymbol{E}$ must be a function of time, as the operator ∇ cannot be. Thus equation (2.12) tells us that a $\boldsymbol{B}$ field which varies in time gives rise in general to an $\boldsymbol{E}$ field which varies both in space and in time. In a similar way, a non-zero $\partial\boldsymbol{E}/\partial t$ gives rise, through equation (2.11), to a spatially and time-varying $\boldsymbol{B}$ field. It is this ability of each of the fields to generate the other which lies at the heart of electromagnetic wave propagation. Our next step is to show rigorously that these coupled fields do indeed propagate through space as electromagnetic waves. Before we do this, however, we should note that Maxwell's equations are linear, that is to say they contain $\boldsymbol{E}$ and $\boldsymbol{B}$ and their derivatives to the first power only. Equations (2.9)–(2.12) are in addition homogeneous, that is to say, as there are no charges or currents present, every term contains either an $\boldsymbol{E}$ or a $\boldsymbol{B}$.

Equations of this sort have the important property that the sum of two different solutions is also a solution. Now we have seen in the previous two sections that equations (2.9)–(2.12) have static field solutions. Thus a general solution to these equations will consist of a sum of static as well as time-varying fields. The physical origin of such static fields would be distributions of static charges and steady currents. In this section, we will assume that such distributions of charges and currents are absent, so there will be no static fields present.

To demonstrate the existence of waves in the electromagnetic field, consider first the $\boldsymbol{E}$ field. We can obtain an equation relating the spatial variations of $\boldsymbol{E}$ to its time variation by eliminating $\boldsymbol{B}$ between the two curl equations (2.11) and (2.12). Taking the curl of equation (2.12) gives

$$\nabla \wedge (\nabla \wedge \boldsymbol{E})=-\frac{\partial}{\partial t}(\nabla \wedge \boldsymbol{B}) \tag{2.13}$$

where we have used the fact that the order of the two operators $\partial/\partial t$ and ∇ can be changed round, as they are independent of each other. Now, substituting from equation (2.11) into the right-hand side of equation (2.13) gives

$$\nabla \wedge (\nabla \wedge \boldsymbol{E}) = -\frac{1}{c^2}\frac{\partial^2 \boldsymbol{E}}{\partial t^2}. \tag{2.14}$$

Using the vector identity

$$\boldsymbol{A} \wedge (\boldsymbol{B} \wedge \boldsymbol{C}) = \boldsymbol{B}(\boldsymbol{A}\cdot\boldsymbol{C}) - \boldsymbol{C}(\boldsymbol{A}\cdot\boldsymbol{B})$$

the double curl on the right-hand side of equation (2.14) becomes

$$\nabla \wedge (\nabla \wedge \boldsymbol{E}) = \nabla(\nabla\cdot\boldsymbol{E}) - \nabla\cdot\nabla\boldsymbol{E}. \tag{2.15}$$

Finally, using $\nabla\cdot\boldsymbol{E}=0$ equation (2.14) can be written

$$\nabla^2\boldsymbol{E} = \frac{1}{c^2}\frac{\partial^2 \boldsymbol{E}}{\partial t^2} \tag{2.16}$$

where, as usual, $\nabla\cdot\nabla$ is written ∇^2.

Similarly, by eliminating $\boldsymbol{E}$ between equations (2.11) and (2.12) we obtain

$$\nabla^2\boldsymbol{B} = \frac{1}{c^2}\frac{\partial^2 \boldsymbol{B}}{\partial t^2}. \tag{2.17}$$

Equations (2.16) and (2.17) are vector wave equations and c is the velocity of propagation of the waves. It should be noted that the displacement current term $(1/c^2)(\partial\boldsymbol{E}/\partial t)$ on the right-hand side of equation (2.11) is crucial to the existence of electromagnetic waves, for without this term Maxwell's equations would not yield wave equations for $\boldsymbol{E}$ and $\boldsymbol{B}$. Notice also that the two vector wave equations are not independent of each other. This is because $\boldsymbol{E}$ and $\boldsymbol{B}$ are related by the two curl equations (2.11) and (2.12). We will employ this connection below. We should remark that the vector wave equation satisfied by $\boldsymbol{E}$ and $\boldsymbol{B}$ describes the propagation of undamped waves. This is just what we should expect as there are no charges or currents present to interact with the fields and hence take energy from them. In the next chapter we will consider the propagation of electromagnetic waves in a conductor and will find that a wave equation describing damped waves applies under these circumstances.

The vector wave equations (2.16) and (2.17) describe the propagation of waves in empty space in full generality; that is, the wavefronts can be plane or spherical or cylindrical or any shape that we care to consider. However, from now on we will confine our attention to *plane waves* as these are the simplest type of wave. Plane waves are also of particular practical interest because a limited portion of a non-planar wavefront which is far from its sources approximates closely to a plane.

We will choose a rectangular cartesian coordinate system because it has the appropriate symmetry, and consider waves that propagate in the positive x direction (see figure 2.6). The two vector wave equations can now be written as a

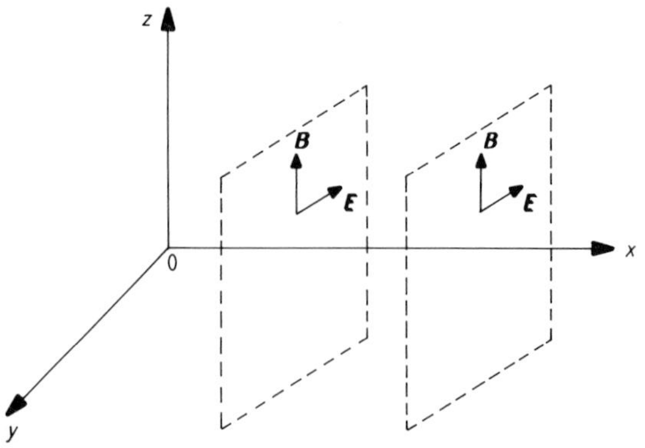

Figure 2.6 A plane wave propagating in the x direction. Two wavefronts are shown. These are planes over which each variable associated with the wave has the same value.

set of six equations in the cartesian components of the two fields. For the x component of the $\boldsymbol{E}$ field we have

$$\nabla^2 E_x = \frac{1}{c^2}\frac{\partial^2 E_x}{\partial t^2} \tag{2.18}$$

with similar equations for the y and z components of the $\boldsymbol{E}$ field and for the x, y and z components of the $\boldsymbol{B}$ field.

Over the unbounded plane wavefronts the magnitudes and directions of the fields will not vary, so the partial derivatives of the field components with respect to y and z will be zero. Thus

$$\begin{aligned}\frac{\partial}{\partial y}\begin{pmatrix}\text{field}\\ \text{component}\end{pmatrix} &= 0\\ \frac{\partial}{\partial z}\begin{pmatrix}\text{field}\\ \text{component}\end{pmatrix} &= 0.\end{aligned} \tag{2.19}$$

The second partial derivatives of the field components with respect to y and z will also be zero, so the three-dimensional wave equations (2.18) reduce to the set of one-dimensional wave equations

$$\begin{aligned}\frac{\partial^2 E_x}{\partial x^2} &= \frac{1}{c^2}\frac{\partial^2 E_x}{\partial t^2}\\ \frac{\partial^2 E_y}{\partial x^2} &= \frac{1}{c^2}\frac{\partial^2 E_y}{\partial t^2}\\ \frac{\partial^2 E_z}{\partial x^2} &= \frac{1}{c^2}\frac{\partial^2 E_z}{\partial t^2}\end{aligned} \tag{2.20}$$

with similar equations for the components of the **B** field. Furthermore, combining equations (2.19) with equation (2.9) gives

$$\frac{\partial E_x}{\partial x}=0. \tag{2.21}$$

On integrating equation (2.21) with respect to x we find that E_x does not vary with position. Using equation (2.19) in the x component of equation (2.11) gives

$$\frac{\partial E_x}{\partial t}=0 \tag{2.22}$$

and integrating this equation with respect to time tells us that E_x does not vary with time either. Therefore, the only solution for E_x is a static uniform field. Earlier on, we set the magnitude of static fields to zero, so that $E_x=0$. Similarly, by substituting equations (2.19) into equations (2.10) and (2.12) we can conclude that $B_x=0$. Thus *plane electromagnetic waves are transverse waves with the electric and magnetic field vectors at right angles to the direction of propagation.*

The next matter we have to decide is the orientation of the **E** and **B** vectors in the xy planes of the wave fronts. Let us suppose that the electric field vector always points along the y direction, that is $E_z=0$. Such a wave is said to be plane-polarised in the y direction. Having fixed the direction of **E** we are not now free to choose the direction of **B**: this is because, as was mentioned above, the **E** and **B** vectors are linked together by the two curl equations (2.11) and (2.12). Substituting the conditions (2.19) into equation (2.12) gives us

$$\frac{\partial E_z}{\partial x}=\frac{\partial B_y}{\partial t} \tag{2.23}$$

$$\frac{\partial E_y}{\partial x}=-\frac{\partial B_z}{\partial t} \tag{2.24}$$

whilst equation (2.11) with equation (2.19) gives

$$-\frac{\partial B_z}{\partial x}=\frac{1}{c^2}\frac{\partial E_y}{\partial t} \tag{2.25}$$

$$\frac{\partial B_y}{\partial x}=\frac{1}{c^2}\frac{\partial E_z}{\partial t}. \tag{2.26}$$

Setting $E_z=0$ in equations (2.23) and (2.26) gives

$$\frac{\partial B_y}{\partial t}=0 \tag{2.27}$$

$$\frac{\partial B_y}{\partial x}=0. \tag{2.28}$$

These latter two equations can be integrated, as were equations (2.21) and (2.22), to yield the solution $B_y=0$. So we conclude that if $E_z=0$, then $B_y=0$. On the other hand, the z component of $\boldsymbol{B}$ is non-zero and can be obtained by solving a one-dimensional wave equation; this follows from equations (2.24) and (2.25). Taking $\partial/\partial t$ of equation (2.24) and $\partial/\partial x$ of equation (2.25), and using the fact that

$$\frac{\partial^2}{\partial x\,\partial t}=\frac{\partial^2}{\partial t\,\partial x}$$

we recover

$$\frac{\partial^2 B_z}{\partial x^2}=\frac{1}{c^2}\frac{\partial^2 B_z}{\partial t^2}. \tag{2.29}$$

So we see that the $\boldsymbol{B}$ vector points entirely in the z direction if the $\boldsymbol{E}$ vector points in the y direction. Thus the $\boldsymbol{E}$ and $\boldsymbol{B}$ vectors in the wavefronts of a plane electromagnetic wave are oriented at right angles to each other (see figure 2.6). If we take $\partial/\partial x$ of equation (2.24) and $\partial/\partial t$ of equation (2.25) we recover the wave equation for E_y:

$$\frac{\partial^2 E_y}{\partial x^2}=\frac{1}{c^2}\frac{\partial^2 E_y}{\partial t^2}. \tag{2.30}$$

It can be verified by substitution that equations (2.29) and (2.30) have the harmonic travelling wave solutions

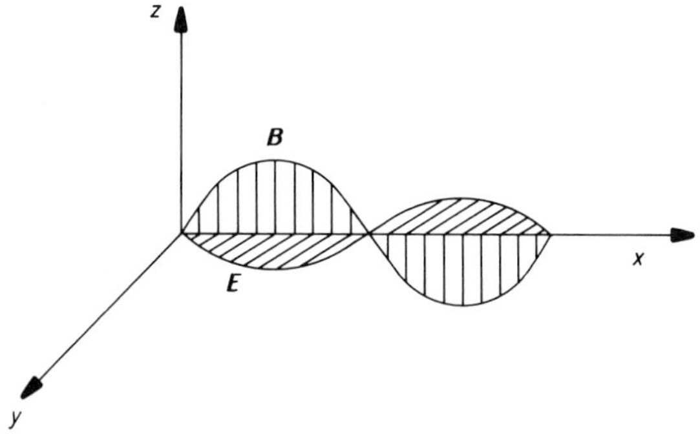

Figure 2.7 A plane-polarised electromagnetic wave at an instant of time, showing the values of the electric vector and the magnetic vector as functions of x; these values are independent of y and z for a plane wave. Note that the $\boldsymbol{E}$ and $\boldsymbol{B}$ vectors are at right angles to each other and in phase. As time passes, the picture moves to the right with the velocity of light. Thus each point in the path of the wave experiences a time-varying electric and magnetic field.

$$B_z = B_0 \sin (2\pi/\lambda)(ct \mp x) \tag{2.31}$$

$$E_y = E_0 \sin (2\pi/\lambda)(ct \mp x). \tag{2.32}$$

This pair of fields represent a plane-polarised electromagnetic wave of wavelength λ travelling with velocity c in the x direction if the negative sign is taken, or in the negative x direction if the positive sign is taken. Notice that the electric and magnetic fields of the wave are in phase with each other, for only then do equations (2.31) and (2.32) satisfy equations (2.24) and (2.25). (See figure 2.7 for an illustration of the relationship of the two fields.) From this substitution we also find that

$$E_0 = cB_0. \tag{2.33}$$

Finally, it should be recalled that, as the Maxwell equations are linear, the principle of superposition applies and more general solutions can be obtained by adding together different plane-polarised electromagnetic wave solutions. An interesting example of such a solution is the case where two plane-polarised electromagnetic waves of the same amplitude, both propagating in the positive x direction, are added together. If one of the waves has its $\boldsymbol{E}$ vector plane-polarised in the z direction whilst the other has its $\boldsymbol{E}$ vector plane-polarised in the y direction, and if there is a phase difference of 90° between the waves, the resultant $\boldsymbol{E}$ vector and $\boldsymbol{B}$ vector have constant amplitude and rotate in the plane of the wave front as the wave advances. (See figure 2.8 for an illustration of the resultant $\boldsymbol{E}$ vector.) Such a wave is described as being circularly polarised.

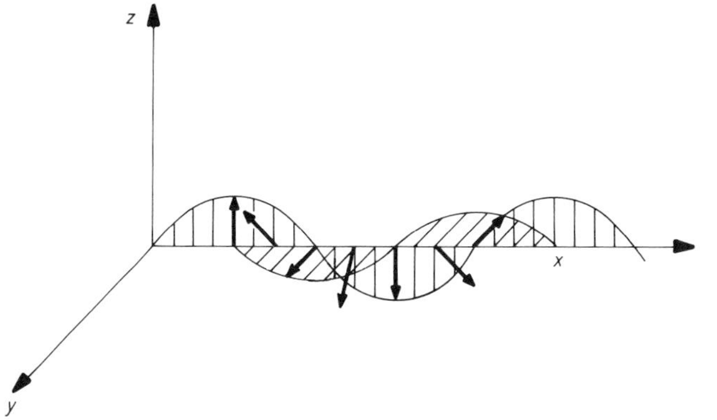

Figure 2.8 Circular polarisation. Two waves of the same λ are travelling along the positive x direction. Only the two electric vectors are shown. The arrows denote the direction of the resultant electric field. A similar construction applies for the two magnetic field vectors. The resultant magnetic field vector is always at right angles to the resultant electric field vector.

2.6 Energy in the Electromagnetic Field

We have seen that electromagnetic waves propagate through space with a speed c. If the waves encounter charges in their path the charges will be influenced by the fields gaining or losing energy, depending on their direction of motion with respect to the $\boldsymbol{E}$ field (see figure 2.9 and below). The law of conservation of total energy leads us to conclude that if, for example, the charges gain energy then this energy must have been present in the wave field at an earlier time. Thus the electromagnetic field has energy associated with it.

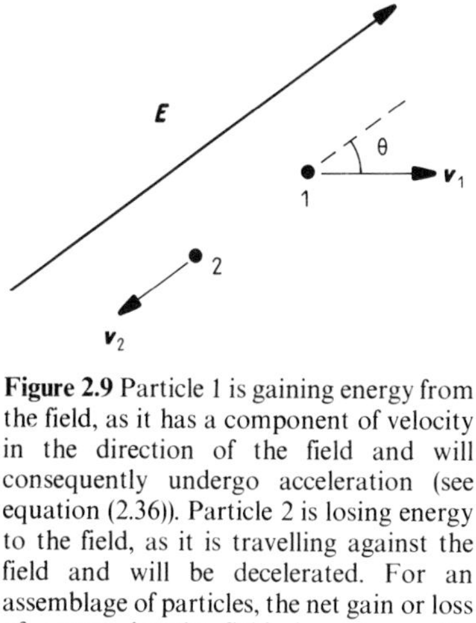

Figure 2.9 Particle 1 is gaining energy from the field, as it has a component of velocity in the direction of the field and will consequently undergo acceleration (see equation (2.36)). Particle 2 is losing energy to the field, as it is travelling against the field and will be decelerated. For an assemblage of particles, the net gain or loss of energy by the field depends on the average motion of the charges with respect to the field.

In this section, we shall find expressions for the energy density and the rate of energy flow in the electromagnetic field. We consider a region of space containing fields and charges of magnitude q. At a point P let the average drift velocity of the charges be $\boldsymbol{v}$ and their density be n per unit volume. Then the current density at P will be

$$\boldsymbol{j}=nq\boldsymbol{v}. \tag{2.34}$$

The force acting on a single particle of charge q and velocity $\boldsymbol{v}_i$ in a region of space containing $\boldsymbol{E}$ and $\boldsymbol{B}$ fields is given by the Lorentz force law (see equation (1.22))

$$\boldsymbol{F}=q(\boldsymbol{E}+\boldsymbol{v}_i \wedge \boldsymbol{B}). \tag{2.35}$$

Therefore, from elementary mechanics, the rate at which the fields do work on this charge is given by

$$\boldsymbol{F}\cdot\boldsymbol{v}_i=q\boldsymbol{E}\cdot\boldsymbol{v}_i+\boldsymbol{v}_i\wedge\boldsymbol{B}\cdot\boldsymbol{v}_i=q\boldsymbol{E}\cdot\boldsymbol{v}_i \tag{2.36}$$

where the fact that $\boldsymbol{v}_i\wedge\boldsymbol{B}\cdot\boldsymbol{v}_i=0$ has been used. (Note that it is a general result that the $\boldsymbol{B}$ field does no work on charged particles as the magnetic force is perpendicular to the velocity of the particles.) Therefore to find the *net* rate at which the fields do work per unit volume on the distribution of charges at P we have to add together the $\boldsymbol{F}\cdot\boldsymbol{v}_i$ for the n particles in a unit volume (see figure 2.9). Doing this gives

$$\begin{pmatrix}\text{rate at which fields}\\ \text{do work per unit}\\ \text{volume at } P\end{pmatrix}=q\sum_{i=1}^{n}\boldsymbol{E}\cdot\boldsymbol{v}_i=q\sum_{i=1}^{n}[E_x(v_i)_x+E_y(v_i)_y+E_z(v_i)_z]. \tag{2.37}$$

Now $(1/n)\sum_{i=1}^{n}(v_i)_x=v_x$, the x component of the average drift velocity of the charges at P. Similarly, $(1/n)\sum_{i=1}^{n}(v_i)_y=v_y$ and $(1/n)\sum_{i=1}^{n}(v_i)_z=v_z$, where v_y and v_z are the y and z components of the average drift velocity of the charges at P. Using these expressions in equation (2.37) we get

$$\begin{pmatrix}\text{rate at which fields}\\ \text{do work per unit}\\ \text{volume at } P\end{pmatrix}=nq\boldsymbol{E}\cdot\boldsymbol{v}=\boldsymbol{E}\cdot\boldsymbol{j} \tag{2.38}$$

where the definition of current density, equation (2.34), has been used. We will now express $\boldsymbol{E}\cdot\boldsymbol{j}$ purely in terms of the fields using the Maxwell equation (1.14) for $\boldsymbol{j}$. We find

$$\boldsymbol{E}\cdot\boldsymbol{j}=\varepsilon_0c^2(\nabla\wedge\boldsymbol{B})\cdot\boldsymbol{E}-\varepsilon_0\boldsymbol{E}\cdot\frac{\partial\boldsymbol{E}}{\partial t}. \tag{2.39}$$

Employing the vector identity

$$\nabla\cdot(\boldsymbol{E}\wedge\boldsymbol{B})=\boldsymbol{B}\cdot(\nabla\wedge\boldsymbol{E})-\boldsymbol{E}\cdot(\nabla\wedge\boldsymbol{B})$$

equation (2.39) becomes

$$\boldsymbol{E}\cdot\boldsymbol{j}=\varepsilon_0c^2\boldsymbol{B}\cdot(\nabla\wedge\boldsymbol{E})-\varepsilon_0c^2\nabla\cdot(\boldsymbol{E}\wedge\boldsymbol{B})-\frac{\varepsilon_0}{2}\frac{\partial}{\partial t}(\boldsymbol{E}\cdot\boldsymbol{E}). \tag{2.40}$$

Finally, using Faraday's law, equation (1.17), in the first term we obtain

$$\boldsymbol{E}\cdot\boldsymbol{j}=-\varepsilon_0c^2\nabla\cdot(\boldsymbol{E}\wedge\boldsymbol{B})-\frac{\varepsilon_0}{2}\frac{\partial}{\partial t}(\boldsymbol{E}\cdot\boldsymbol{E}+c^2\boldsymbol{B}\cdot\boldsymbol{B}). \tag{2.41}$$

If there are no currents present, $\boldsymbol{E}\cdot\boldsymbol{j}=0$, so there is no conversion of field energy to other forms of energy and equation (2.41) has exactly the same form as equation (1.12) which expresses the conservation of electric charge. This suggests that we interpret equation (2.41) as the equation of energy conservation in the electromagnetic field. Then the quantity

$$\boldsymbol{S}=\varepsilon_0c^2\boldsymbol{E}\wedge\boldsymbol{B} \tag{2.42}$$

is the rate of flow of energy per unit area in the direction of the waves for plane waves and the quantity

$$U=(\varepsilon_0/2)(\boldsymbol{E}\cdot\boldsymbol{E}+c^2\boldsymbol{B}\cdot\boldsymbol{B}) \tag{2.43}$$

is the energy density in the field. The vector $\boldsymbol{S}$ is known as the Poynting vector. Rewriting equation (2.41) in terms of $\boldsymbol{S}$ and U we obtain

$$\frac{\partial U}{\partial t}=-\nabla\cdot\boldsymbol{S}-\boldsymbol{E}\cdot\boldsymbol{j}. \tag{2.44}$$

In words this equation tells us that

$$\left(\begin{array}{l}\text{The rate at which field energy}\\ \text{is changing per unit volume}\end{array}\right)$$

$$=-\left(\begin{array}{l}\text{(The net rate at which energy is flowing out of}\\ \text{a region per unit volume)}+\text{(the rate at which}\\ \text{the field is losing energy per unit volume by}\\ \text{doing work on charges)}\end{array}\right).$$

The minus sign tells us that when either of the two terms on the right-hand side is positive, the term causes a decrease in the field energy.

Let us now apply equations (2.42) and (2.43) to the case of plane waves. From equation (2.33), $E=cB$, so

$$S=\varepsilon_0 cE^2 \tag{2.45}$$

in a direction normal to the wave front and

$$U=\varepsilon_0 E^2. \tag{2.46}$$

We can also obtain the energy density from equation (2.45) because the energy that crosses a unit area in 1 s occupies a volume c. Thus the energy per unit volume is $\varepsilon_0 E^2$ which agrees with equation (2.46). So the two expressions (2.45) and (2.46) give consistent results for plane waves.

Finally we will show that for a plane electromagnetic wave propagating in a vacuum the energy density associated with the electric field is equal to the energy density associated with the magnetic field. These energy densities are, from equation (2.43), respectively

$$U_{\text{elect}}=\tfrac{1}{2}\varepsilon_0 E^2 \qquad U_{\text{mag}}=\tfrac{1}{2}\varepsilon_0 c^2 B^2.$$

The magnitudes of E and B are related by equation (2.33) so evidently

$$U_{\text{elect}}=U_{\text{mag}}.$$

Maxwell's Equations and Matter

3

3.1 The Macroscopic Maxwell Equations

The subject we are going to consider now is the calculation of fields inside matter. We have seen in the previous chapter how Maxwell's equations (1.5), (1.7), (1.14) and (1.17) enable us to calculate the $\boldsymbol{E}$ and $\boldsymbol{B}$ fields once the individual sources of the fields are specified. We have also seen that, in regions of space which contain no charges and currents, wave solutions for $\boldsymbol{E}$ and $\boldsymbol{B}$ exist. Maxwell's equations in the form given in the box at the end of Chapter 1 are known as the 'microscopic vacuum field equations'. 'Microscopic' because the fields are specified down to scales of atomic size and 'vacuum' because the charges and fields reside in a vacuum. In principle, we could use the microscopic vacuum field equations to calculate the $\boldsymbol{E}$ and $\boldsymbol{B}$ fields inside matter. If we were to do this, we would have to treat the matter as a collection of charges in vacuum. Then *all* these charges would have to be put into the right-hand side of equation (1.5). Correspondingly, all sources of current would need to be included in the right-hand side of equation (1.14). Clearly, such a calculation, which requires the detailed specification of all the myriads of sources that are present in bulk matter, is totally impracticable. However, it is not necessary to know the $\boldsymbol{E}$ and $\boldsymbol{B}$ fields in the microscopic detail that such a calculation would yield. What we do require is a knowledge of the fields averaged over dimensions which are large on the atomic scale.

These averaged $\boldsymbol{E}$ and $\boldsymbol{B}$ fields can be obtained by solving the 'macroscopic field equations' which are

$$\nabla \cdot \boldsymbol{D} = \rho_{\mathrm{f}} \tag{3.1}$$

$$\nabla \cdot \boldsymbol{B} = 0 \tag{3.2}$$

$$\nabla \wedge \boldsymbol{H} = \boldsymbol{j}_{\mathrm{f}} + \frac{\partial \boldsymbol{D}}{\partial t} \tag{3.3}$$

$$\nabla \wedge \boldsymbol{E} = -\frac{\partial \boldsymbol{B}}{\partial t}. \tag{3.4}$$

In addition, we need to know the constitutive relations which connect the auxiliary fields $\boldsymbol{D}$ and $\boldsymbol{H}$ to the fundamental fields $\boldsymbol{E}$ and $\boldsymbol{B}$. Thus

$$\boldsymbol{D}=f(\boldsymbol{E}, \boldsymbol{B}) \tag{3.5}$$

$$\boldsymbol{H}=f(\boldsymbol{E}, \boldsymbol{B}). \tag{3.6}$$

A full justification of equations (3.1) to (3.6) cannot be undertaken here, but we can get some understanding of how these equations are established from the following remarks. Consider what happens when an electrostatic field is applied to an insulating material. The insulator atoms will become polarised, that is, the centres of positive and negative charge will be displaced in opposite directions. This polarisation will, if the material is isotropic, be parallel to the $\boldsymbol{E}$ field and will normally be proportional to the strength of $\boldsymbol{E}$. A polarisation vector $\boldsymbol{P}$ can be defined within the material. The magnitude of $\boldsymbol{P}$ at each point is the dipole moment per unit volume there and its direction is that of the polarisation at the point. Note that the magnitude of $\boldsymbol{P}$ at a point is obtained by averaging over a volume that is large on the atomic scale, so $\boldsymbol{P}$ is a macroscopic vector. Figure 3.1 shows the effect of a uniform electric field on a slab of *homogeneous* material. Inside the material, the relative movement of charge centres does not cause any volume distributions of polarisation charge to appear because charges moving out of any volume are compensated by an equal amount of charge of the same sign moving into the volume. However, at the surfaces of the material, as figure 3.1 shows, net polarisation charges will appear. Note that in the more general case of an *inhomogeneous* material there will be volume as well as surface polarisation charges.

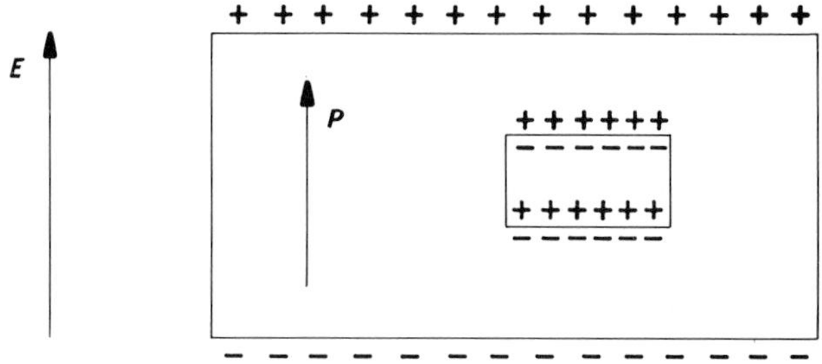

Figure 3.1

Now the polarisation charges will act as sources of the $\boldsymbol{E}$ field, along with other sources of charge which are present, and so should be included on the right-hand side of the first Maxwell equation $\nabla \cdot \boldsymbol{E}=\rho/\varepsilon_0$. To distinguish polarisation charge from other sorts of charge we will write this equation in the form

$$\nabla \cdot \boldsymbol{E}=\rho_{\mathrm{f}}+\rho_{\mathrm{pol}} \tag{3.7}$$

where ρ_{pol} denotes polarisation charge density and ρ_f is called the 'free charge' density; this latter comprises all sources of charge other than ρ_{pol}. Now, ρ_{pol} can be obtained from $\boldsymbol{P}$ as follows. Lines of $\boldsymbol{P}$ start on *negative* polarisation charge and end on *positive* polarisation charge (see figure 3.1). Thus, by analogy with Gauss's flux law, equation (1.3) for $\boldsymbol{E}$, we can write a corresponding flux law for $\boldsymbol{P}$

$$\int_A \boldsymbol{P}\cdot \mathrm{d}\boldsymbol{A} = -\int_V \rho_{pol}\,\mathrm{d}V$$

or in differential form

$$\nabla\cdot\boldsymbol{P} = -\rho_{pol}.$$

The negative sign is needed because lines of $\boldsymbol{P}$ run in the opposite sense to lines of $\boldsymbol{E}$. Now we substitute this result into equation (3.7), and obtain

$$\nabla\cdot\boldsymbol{E} = (\rho_f - \nabla\cdot\boldsymbol{P})/\varepsilon_0$$

or

$$\nabla\cdot(\varepsilon_0\boldsymbol{E} + \boldsymbol{P}) = \rho_f.$$

Note that as $\boldsymbol{P}$ is a macroscopic quantity, then so will be the $\boldsymbol{E}$ vector obtained from this equation. Finally, the auxiliary vector $\boldsymbol{D} = \varepsilon_0\boldsymbol{E} + \boldsymbol{P}$ is introduced and the first of the macroscopic Maxwell equations

$$\nabla\cdot\boldsymbol{D} = \rho_f \tag{3.1}$$

is obtained.

For the special case of a homogeneous isotropic medium, $\boldsymbol{P}$ is related to $\boldsymbol{E}$ by

$$\boldsymbol{P} = \varepsilon_0\chi\boldsymbol{E}$$

where χ is a constant. In this case, equation (3.5) takes the simple form

$$\boldsymbol{D} = \varepsilon_0(1+\chi)\boldsymbol{E} = \varepsilon\boldsymbol{E} \tag{3.8}$$

where ε is a constant called the permittivity of the medium.

In a similar way, a magnetic field $\boldsymbol{B}$ creates a magnetisation vector $\boldsymbol{M}$ inside the material and an analogous argument to the one above would lead us to equation (3.3), where $\boldsymbol{j}_f$ denotes currents arising from the motion of free charge. The auxiliary vector $\boldsymbol{H}$ is defined by

$$\boldsymbol{H} = \boldsymbol{B}/\mu_0 - \boldsymbol{M}.$$

For the case of a homogeneous isotropic medium we obtain

$$\boldsymbol{M} = \chi_m\boldsymbol{H}$$

where χ_m is a constant. Then

$$\boldsymbol{H} = \boldsymbol{B}/\mu_0 - \chi_m\boldsymbol{H}$$

so the constitutive relation for $\boldsymbol{H}$ is

$$H = B/\mu_0(1+\chi_m) = B/\mu \tag{3.9}$$

where μ is called the permeability of the medium.

3.2 Electromagnetic Waves in a Homogeneous and Isotropic Conductor

We will now use the equations introduced in the previous section to study the propagation of electromagnetic waves in an infinite homogeneous and isotropic conductor whose conductivity obeys Ohm's law. We have specified an infinite conductor, so that we do not have to consider reflections from boundaries. These are the same boundary conditions we adopted in Chapter 2, when considering waves in a vacuum.

In a conductor, there are free charges present—both the negatively charged conduction electrons and the positively charged ions from which the electrons have come. As there are equal amounts of both sorts of charge, the net charge density at any point will be zero i.e. $\rho_f = 0$. Also, for a conductor obeying Ohm's law, we can write (see §1.7)

$$j_f = \sigma E \tag{3.10}$$

where σ is the conductivity.

When these values for ρ_f and j_f and the relationships $D = \varepsilon E$ and $H = B/\mu$ are inserted into equations (3.1)–(3.4), we obtain

$$\nabla \cdot (\varepsilon E) = 0 \tag{3.11}$$

$$\nabla \cdot B = 0 \tag{3.12}$$

$$\nabla \wedge (B/\mu) = \sigma E + \frac{\partial}{\partial t}(\varepsilon E) \tag{3.13}$$

$$\nabla \wedge E = -\frac{\partial B}{\partial t}. \tag{3.14}$$

The next step is to proceed just as we did in Chapter 2. We eliminate B between the two curl equations (3.13) and (3.14) and obtain the following equation for E

$$\nabla^2 E = \mu\sigma \frac{\partial E}{\partial t} + \mu\varepsilon \frac{\partial^2 E}{\partial t^2}. \tag{3.15}$$

Similarly, the equation for B is found to be

$$\nabla^2 B = \mu\sigma \frac{\partial B}{\partial t} + \mu\varepsilon \frac{\partial^2 B}{\partial t^2}. \tag{3.16}$$

These two equations resemble the wave equations (2.16) and (2.17) that we obtained in Chapter 2, except that they have an extra term on their right-hand sides. Notwithstanding this difference, they are clearly some sort of wave equation whose exact nature we now have to determine. Let us therefore, as before, specialise to the case of plane electromagnetic waves propagating in the positive x direction. The reasoning of Chapter 2 leads us to conclude again that $E_x = B_x = 0$; thus we are dealing once more with transverse waves. The remaining components of the $\boldsymbol{E}$ and $\boldsymbol{B}$ vectors are related by the two curl equations (3.13) and (3.14), which give in component form

$$\frac{\partial E_z}{\partial x} = \frac{\partial B_y}{\partial t} \tag{3.17}$$

$$\frac{\partial E_y}{\partial x} = -\frac{\partial B_z}{\partial t} \tag{3.18}$$

$$-\frac{\partial B_z}{\partial x} = \mu\sigma E_y + \mu\varepsilon \frac{\partial E_y}{\partial t} \tag{3.19}$$

$$\frac{\partial B_y}{\partial x} = \mu\sigma E_z + \mu\varepsilon \frac{\partial E_z}{\partial t}. \tag{3.20}$$

If, as before, we assume that $E_z = 0$, then $\partial B_y/\partial t = \partial B_y/\partial x = 0$, which gives $B_y = 0$. So once again, if $E_z = 0$, then $B_y = 0$, and the $\boldsymbol{E}$ and $\boldsymbol{B}$ vectors are perpendicular to each other in the wavefronts. Thus from equation (3.15), with $E_x = E_z = 0$, we obtain the following equation for E_y

$$\frac{\partial^2 E_y}{\partial x^2} = \mu\sigma \frac{\partial E_y}{\partial t} + \mu\varepsilon \frac{\partial^2 E_y}{\partial t^2} \tag{3.21}$$

with a similar equation for B_z. Let us now look into the purpose of the term $\mu\sigma\, \partial E_y/\partial t$ which distinguishes equation (3.21) from the simple wave equation (2.30) of the previous chapter. This extra term disappears when $\sigma = 0$, so is evidently something to do with the finite conductivity of the medium. Now, the effect of this conductivity is to cause conduction currents to flow under the influence of the electric vector of the wave. Furthermore, these currents will be dissipated irreversibly through Ohmic heating, so there must be a loss of energy from the wave as it propagates. Evidently, then, equation (3.21) describes the propagation of damped waves and $\mu\sigma\, \partial E_y/\partial t$ is a damping term. We will therefore try as a solution of equation (3.21)

$$E_y = E_0\, \mathrm{e}^{-\gamma x} \sin(\omega t - kx) \tag{3.22}$$

where γ is real and positive.

This expression should be compared with equation (2.32); the essential difference is that here the amplitude of the wave, $E_0\, \mathrm{e}^{-\gamma x}$, decays exponentially with distance. Note that in equation (3.22) k is written for $2\pi/\lambda$, the wave number, and ω, the angular frequency, is written for kv where v is the velocity of the wave through the conductor.

On substituting equation (3.22) into equation (3.21) we obtain

$$(\gamma^2 - k^2 + \mu\varepsilon\omega^2)\sin(\omega t - kx) + (2\gamma k - \mu\sigma\omega)\cos(\omega t - kx) = 0.$$

The only way that this equation can be satisfied is for the coefficients of the sine and cosine terms to be independently equal to zero. Thus we find

$$\gamma^2 - k^2 + \mu\varepsilon\omega^2 = 0$$

and

$$2\gamma k - \mu\sigma\omega = 0.$$

Eliminating k between these equations and solving the resultant quadratic equation for γ^2 gives

$$\gamma^2 = -\tfrac{1}{2}\mu\varepsilon\omega^2 \pm \tfrac{1}{2}\mu\sigma\omega(1 + \varepsilon^2\omega^2/\sigma^2)^{1/2}. \tag{3.23}$$

As γ is real we must take the positive sign; so, after some rearrangement

$$\gamma^2 = \tfrac{1}{2}\mu\sigma\omega[(1 + \varepsilon^2\omega^2/\sigma^2)^{1/2} - \varepsilon\omega/\sigma]. \tag{3.24}$$

Thus we have shown that equation (3.22) is a solution of equation (3.21) with γ given by equation (3.24).

For a good conductor, ε/σ is very small, e.g. for copper $\sigma = 5.8 \times 10^7\ \mathrm{S\,m^{-1}}$ and $\varepsilon \sim \varepsilon_0 = 8.85 \times 10^{-12}\ \mathrm{F\,m^{-1}}$, so that $\varepsilon/\sigma \simeq 1.5 \times 10^{-19}\ \mathrm{s}$. So, for angular frequencies $\omega \ll 10^{19}\ \mathrm{rad\,s^{-1}}$, the quantity $\varepsilon\omega/\sigma$ is very small. In this case the quantity in square brackets in equation (3.24) is approximately unity; thus

$$\gamma = (\mu\sigma\omega/2)^{1/2}$$

and hence

$$k = (\mu\sigma\omega/2)^{1/2}. \tag{3.25}$$

Equation (3.22) tells us that the electric field vector is proportional to $\exp-(\gamma x)$, thus its amplitude will decay by a factor $1/\mathrm{e}$ in a distance $\delta = 1/\gamma$; so for a good conductor we have

$$\delta = (2/\mu\sigma\omega)^{1/2}$$

where δ is called the skin depth. For copper, when the frequency of the radiation is 50 Hz, $\delta = 0.9$ cm, whilst at a frequency of 5×10^{10} Hz, $\delta \simeq 3 \times 10^{-7}$ m.

Let us turn now to the magnetic field. The equation for B_z which goes with equation (3.21) for E_y is

$$\frac{\partial^2 B_z}{\partial x^2} = \mu\sigma\frac{\partial E_z}{\partial t} + \mu\varepsilon\frac{\partial^2 E_z}{\partial t^2}. \tag{3.26}$$

As this equation is identical to equation (3.21), we are tempted to write down as its solution $B_z = B_0\,\mathrm{e}^{-\gamma x}\sin(\omega t - kx)$. However, we must not forget that E_y and B_z are not independent of each other, but are related via the curl equations (3.17)–(3.20). If we substitute E_y from equation (3.22) and the similar expression for B_z above into equation (3.18) we obtain

$$-E_0\gamma\sin(\omega t - kx) = (E_0 k - \omega B_0)\cos(\omega t - kx).$$

The only solution to this equation is the trivial one $E_0 = B_0 = 0$. Evidently the $\boldsymbol{B}$ vector and the $\boldsymbol{E}$ vector must be out of phase, so we must write

$$B_z = B_0 \, e^{-\gamma x} \sin(\omega t - kx - \phi). \tag{3.27}$$

Now on substitution of B_z from (3.27) and E_y for (3.22) into (3.18) we get, on equating the coefficients of $\sin(\omega t - kx)$ and $\cos(\omega t - kx)$, respectively,

$$E_0 \gamma = B_0 \omega \sin \phi$$

and

$$kE_0 = B_0 \omega \cos \phi$$

Hence,

$$E_0 = vB_0 \cos \phi \tag{3.28}$$

as $\omega / k = v$, the velocity of the wave, and

$$\tan \phi = \gamma / k.$$

When equations (3.25) hold, $\tan \phi = 1$. So, for radio waves and microwaves, the E vector leads the B vector by 45°.

We have shown that an electromagnetic wave propagates in a conductor with a speed $v = \omega / k$ where k is given by equation (3.25), therefore

$$v = \left(\frac{2\omega}{\mu \sigma} \right)^{1/2}.$$

Thus the speed of the wave is proportional to the square root of its frequency. For example the velocity of 3 GHz microwaves in copper is

$$v = \left(\frac{2 \times 2\pi \times 3 \times 10^9}{4\pi \times 10^{-7} \times 5.8 \times 10^7} \right)^{1/2} = 2.27 \times 10^4 \text{ m s}^{-1}.$$

This behaviour should be compared with that of electromagnetic waves propagating in a vacuum where the velocity is independent of frequency and always has the value $c = 2.99 \times 10^8 \text{ m s}^{-1}$.

Finally we will show that, for an electromagnetic wave propagating in a good conductor, nearly all of the wave's energy is carried by the magnetic component. Equations (2.42) and (2.43) of §2.6 define, respectively, the Poynting vector and the energy density of an electromagnetic wave propagating in a vacuum. We can obtain the corresponding expressions for waves propagating in matter by retracing the steps of §2.6, but using the macroscopic field equations of §3.1 in place of the vacuum field equations (1.5), (1.7), (1.14) and (1.17). When this is done, in the case of an isotropic homogeneous medium, the Poynting vector becomes

$$\boldsymbol{S} = \boldsymbol{E} \wedge \boldsymbol{H}$$

and the energy density becomes

$$U = \tfrac{1}{2}(\varepsilon E^2 + \mu H^2).$$

Note that these expressions reduce to those obtained previously when $\varepsilon = \varepsilon_0$, $\mu = \mu_0$ and $\boldsymbol{H} = \boldsymbol{B}/\mu_0$.

The ratio of electric to magnetic energy density is therefore

$$\frac{\frac{1}{2}\varepsilon E_0^2}{\frac{1}{2}\mu H_0^2}.$$

Now from equations (3.28) and (3.9)

$$H_0 = \frac{E_0}{\mu v \cos\phi}$$

where $v = \omega/k$, k is given by equation (3.25) and for a good conductor $\cos\phi = 1/\sqrt{2}$ as shown above. Thus

$$\frac{\frac{1}{2}\varepsilon E_0^2}{\frac{1}{2}\mu H_0^2} = \frac{\varepsilon\omega}{\sigma}.$$

We pointed out above that $\varepsilon\omega/\sigma$ is a very small quantity for a good conductor for all frequencies up to and including optical frequencies. Thus to a very good approximation the energy is located in the magnetic component of a wave propagating in a conductor. This result should be compared with that obtained at the end of §2.6. There we showed that when an electromagnetic wave propagates through a vacuum its energy is shared equally between the electric and magnetic components.

3.3 Radiation Pressure

As we have seen, an electromagnetic wave which penetrates a conductor produces a flow of current in it. If the wave is plane and moving in the x direction and the electric vector is in the y direction, then the current density is given by $j_y = \sigma E_y$ (as shown in figure 3.2). The magnetic vector of the wave will exert a force on this current corresponding to $\boldsymbol{j} \wedge \boldsymbol{B}$ per unit volume. In the case illustrated, this force is $j_y B_z$; the vector product implies that it acts in the positive x direction. The force $\mathrm{d}F$ on an element of volume $\mathrm{d}x\mathrm{d}y\mathrm{d}z$ of the conductor therefore amounts to

$$\mathrm{d}F = j_y \mu_0 H_z \,\mathrm{d}x\mathrm{d}y\mathrm{d}z$$

and the corresponding pressure $\mathrm{d}P$ is

$$\mathrm{d}P = \frac{\mathrm{d}E}{\mathrm{d}x\mathrm{d}y} = j_y \mu_0 H_z \,\mathrm{d}z.$$

In a conductor the ratio of displacement current to conduction current is

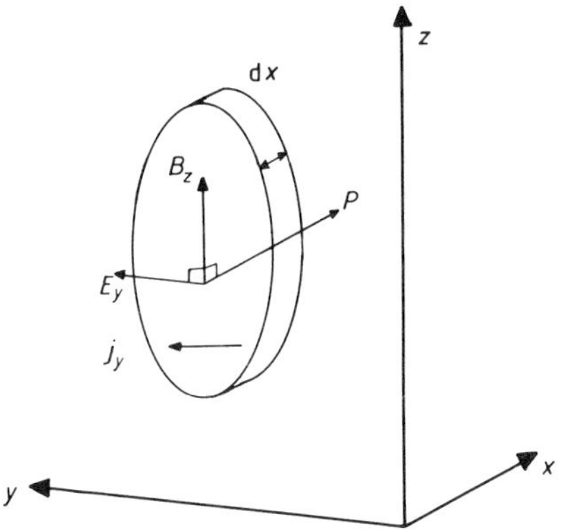

Figure 3.2

$$\left(\frac{\partial(\varepsilon E)}{\partial t}\right)\bigg/ j = \left(\frac{\partial(\varepsilon E)}{\partial t}\right)\bigg/ \sigma E = \frac{\varepsilon\omega}{\sigma}. \tag{3.29}$$

We noted in §3.2 that $\varepsilon\omega/\sigma$ is very small for a good conductor over a wide range of frequencies, so we may therefore write equation (3.13) as $\nabla \wedge \boldsymbol{H} = \boldsymbol{j}$ which gives (in this instance) $j_y = -\partial H_z/\partial x$, because the wave is transverse.

In §3.2 of this chapter we saw that an electromagnetic wave dies away rapidly in a conductor. It is therefore permissible to write that $H_z \to 0$ as $x \to \infty$. Inserting these limits into equation (3.29) gives

$$P = \mu_0 \int_0^\infty j_y H_z \, dx = -\mu_0 \int_0^\infty H_z \frac{\partial H_z}{\partial x} dx = \tfrac{1}{2}\mu_0 H^2 \tag{3.30}$$

where H is the instantaneous value of H_z at the surface of the conductor. We now need to relate this H at the surface of the conductor to H_i, the instantaneous value of the $\boldsymbol{H}$ vector in the incident electromagnetic wave. Conductors are very good reflectors of electromagnetic radiation, so there will be a reflected wave whose intensity is very nearly equal to that of the incident wave. The magnetic vectors of the incident and reflected waves are in the same direction, so the requirement that $\boldsymbol{H}$ must have the same value on both sides of the boundary means that $H = 2H_i$. The electric vectors of the incident and reflected waves are in opposite directions: they therefore almost cancel, leaving only a very small $\boldsymbol{E}$ vector on the conductor side of the boundary (see figure 3.3). This arrangement of $\boldsymbol{E}$ and $\boldsymbol{H}$ vectors at the boundary corresponds to the fact that the ratio of H/E must increase on going from a vacuum to a conductor, as can be seen by comparing equation (3.28) with equation (2.33). The pressure can therefore be written in terms of the instantaneous value of the $\boldsymbol{H}$ vector of the incident wave as

$$P = 2\mu_0 H_i^2.$$

We see from equations (2.43) and (2.33), putting $B_i = \mu_0 H_i$, that the energy density in the incident beam of radiation can be written

$$U = \mu_0 H_i^2.$$

The radiation pressure on the conductor in terms of U is therefore

$$P = 2U.$$

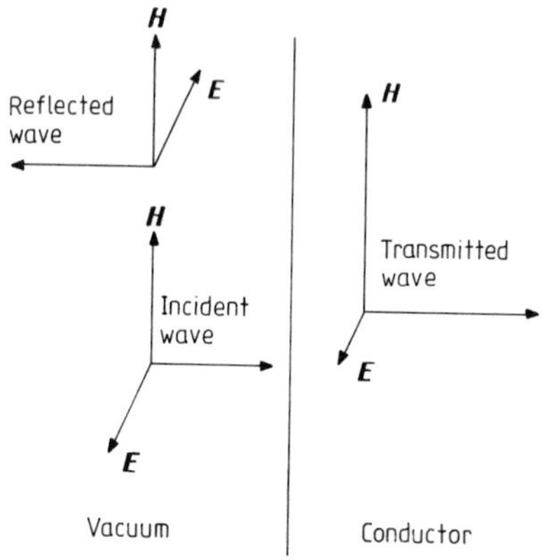

Figure 3.3

We can see from dimensional considerations that P is also the momentum per unit area per second transferred to the conductor by the beam of radiation. The factor two appears in the above expression because the incident beam is almost all reflected back from the conductor's surface, and so has its direction of propagation reversed. The conservation of linear momentum tells us that this reversing of the beam doubles the momentum imparted to the conductor over the value for a beam of radiation that is simply absorbed by the surface. From this we see that U must also be momentum per second per unit area carried by the beam. But from equations (2.45) and (2.46), the beam S crossing unit area per unit time is related to U, regarded as the energy density, by

$$U = S/c$$

since the beam is travelling at the speed of light. So we seen now that this is also the relationship between energy and momentum carried by electromagnetic waves in a vacuum. Note that this is the same as the relationship that appears in the photon picture of light.

If the radiation is diffuse (i.e. moving in all directions), we can imagine that it consists of plane waves moving in each of the x, y and z directions. If the radiation is isotropic, that is it has the same intensity in all directions, then these components contribute to the energy density equally, but only the radiation in the x direction will contribute to the pressure on the conductor. For such radiation we can therefore argue from symmetry that the pressure on the conductor is

$$P=\tfrac{2}{3}U.$$

Note that when radiation impinges on a dielectric material which absorbs it then the pressure is given by

$$P=U$$

for directed radiation and by

$$P=\tfrac{1}{3}U$$

for diffuse radiation. As no currents flow in dielectric materials, the mechanism by which the wave imparts a force is quite different from the mechanism described here for a conductor.

As an example of a situation where radiation pressure has to be taken into account, we will calculate the force exerted by solar radiation on an artificial satellite in orbit about the Earth. At the distance of the Earth from the Sun, the Poynting vector has a magnitude of $1.4\,\mathrm{kW\,m^{-2}}$. The pressure on a metal satellite will therefore be

$$\frac{2S}{c}=\frac{2\times1.4\times10^{3}}{3\times10^{8}}=9.3\times10^{-6}\,\mathrm{N\,m^{-2}}.$$

The effect of this pressure will depend on the surface area of the satellite and on its mass. An extreme case is the balloon satellite Echo I, which was launched in 1960. It was a metallised sphere of 34 metres diameter and had a mass m of 70.5 kg. Thus for Echo I the total force due to solar radiation was

$$9.3\times10^{-6}\times\pi\left(\frac{34}{2}\right)^{2}=8.4\times10^{-3}\,\mathrm{N}$$

and the gravitational force at a height of 320 km above the Earth's surface was

$$F=\frac{GMm}{(a+320)^{2}}=\frac{GMm}{a^{2}}\left(1-\frac{640}{a}\right)$$

where a, the radius of the Earth, is 6400 km and GM/a^{2} the acceleration due to gravity at the Earth's surface is $9.8\,\mathrm{m\,s^{-2}}$, so

$$F=9.8\times70.5\times0.9=622\,\mathrm{N}.$$

The ratio of solar radiation force to gravitational force is therefore

$$\frac{8.4\times10^{-3}}{622}=1.35\times10^{-5}.$$

Despite the small size of this ratio the effect of the radiation force was to perturb the satellite's orbit by several kilometres.

3.4 Diffusion of a Magnetic Field

We finally turn to a problem which requires us to use most of the fundamental equations introduced in the first chapter. This relates to the way in which the strength of a magnetic field in a conductor changes with time. We begin with equation (1.14). Assuming that the electric field does not change with time (i.e. $\partial \boldsymbol{E}/\partial t=0$), this becomes

$$\nabla \wedge \boldsymbol{B}=\mu \boldsymbol{j}.$$

We now substitute for $\boldsymbol{j}$ from equation (1.23), namely

$$\boldsymbol{j}=\sigma(\boldsymbol{E}+\boldsymbol{v} \wedge \boldsymbol{B})$$

giving

$$\nabla \wedge \boldsymbol{B}=\sigma\mu(\boldsymbol{E}+\boldsymbol{v} \wedge \boldsymbol{B}).$$

Next we multiply both sides of this equation by the curl operator:

$$\nabla \wedge \nabla \wedge \boldsymbol{B}=\sigma\mu[\nabla \wedge \boldsymbol{E}+\nabla \wedge (\boldsymbol{v} \wedge \boldsymbol{B})]. \tag{3.31}$$

As we saw in Chapter 2, the left-hand side can be replaced using the vector identity

$$\nabla \wedge \nabla \wedge \boldsymbol{B}=\nabla(\nabla \cdot \boldsymbol{B})-\nabla^2.$$

However, equation (1.7) tells us that $\nabla \cdot \boldsymbol{B}=0$, so the left-hand side of equation (3.31) becomes

$$\nabla \wedge \nabla \wedge \boldsymbol{B}=-\nabla^2 \boldsymbol{B}.$$

But we also know from equation (1.17) that

$$\nabla \wedge \boldsymbol{E}=-\frac{\partial \boldsymbol{B}}{\partial t}.$$

Substituting for $\nabla \wedge \nabla \wedge \boldsymbol{B}$ and $\nabla \wedge \boldsymbol{E}$ in (3.31), we find that

$$-\nabla^2 \boldsymbol{B}=\sigma\mu\left[-\frac{\partial \boldsymbol{B}}{\partial t}+\nabla \wedge (\boldsymbol{v} \wedge \boldsymbol{B})\right]$$

or, rearranging,

$$\frac{\partial \boldsymbol{B}}{\partial t}=\nabla \wedge (\boldsymbol{v} \wedge \boldsymbol{B})+\eta\nabla^2 \boldsymbol{B} \tag{3.32}$$

where $\eta=1/\mu\sigma$. This equation gives the change in $\boldsymbol{B}$ with time.

The velocity $\boldsymbol{v}$ in equation (3.32) represents the motion of the body. If we take a stationary conductor—say a copper sphere sitting on a laboratory table—then $\boldsymbol{v}$ is obviously zero. The equation then reduces to

$$\frac{\partial \boldsymbol{B}}{\partial t}=\eta\nabla^2 \boldsymbol{B}. \tag{3.33}$$

This is one of the commonest differential equations in physics; it represents diffusion. It is usually applied to the diffusion of one substance through another. For example, if a lump of sugar is dropped into a cup of coffee, then this equation will represent the rate at which the sugar spreads out through the coffee. In this case, however, it represents the rate of diffusion of the magnetic field. A magnetic field is introduced into a conducting body and the source of the field is then removed. The field will gradually disappear. The speed with which this occurs will depend on the constant η—that is, on the conductivity σ. If the conductivity is very high, we can write $\sigma \rightarrow \infty$. This implies that the right-hand side of equation (3.33) tends to 0, or $\partial \boldsymbol{B}/\partial t \rightarrow 0$. In other words, $\boldsymbol{B}$ does not change with time; the magnetic flux is trapped within the conductor. As the conductivity decreases, so the speed of diffusion increases. For a copper sphere of one metre radius, the disappearance of a magnetic field takes less than ten seconds.

We can understand what is happening in qualitative terms by invoking one of the properties of lines of force that we discussed in Chapter 1. We noted there that such lines tend to repel each other at right angles to their lengths. Figure 3.4 shows a conducting sphere containing a magnetic field. It is obvious that mutual repulsion will lead to the ejection of lines of force sideways, as indicated by the arrows. The speed with which the field disappears will depend on the resistance of the sphere to the motion of the lines of force. This depends, in turn, on the conductivity of the sphere.

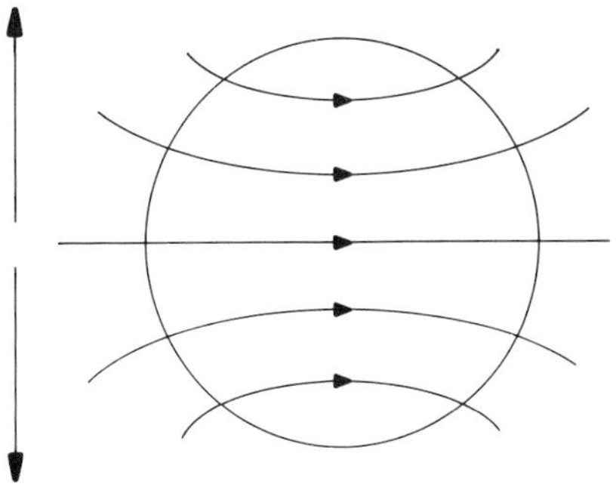

Figure 3.4

Another way of looking at this is to note that the changing magnetic flux creates a flow of current in the sphere. As we saw in Chapter 1, this current will act to maintain the field. The higher the conductivity of the sphere, the longer the current will flow, and so the longer the magnetic field will take to disappear.

3.5 Conclusion

We have followed through the applications of Maxwell's equations to the point where they are actually being used in current research. Thus equation (3.32) is the basic equation at the centre of contemporary discussions concerning the origin of the Earth's magnetic field.

Some general remarks about applying Maxwell's equations can be made in conclusion. The solution of problems with these equations normally follows a logical sequence, which can be summarised as follows.

1 Decide which of the equations is required for the problem concerned.
2 Eliminate quantities that are not required—if possible, from the vector form of the equations.
3 To obtain detailed answers, especially if numerical values are demanded, generally requires conversion of the vector equations to component form.

A glance back through the applications presented in this book should show these steps in operation.

Index